The Mass Gap and its Applications

The Mass Gap and its Applications

VAKHTANG GOGOKHIA
GERGELY GÁBOR BARNAFÖLDI

Institute for Particle & Nuclear Physics,
Wigner Research Centre for Physics,
Hungarian Academy of Sciences, Hungary

NEW JERSEY • LONDON • SINGAPORE • BEIJING • SHANGHAI • HONG KONG • TAIPEI • CHENNAI

Published by

World Scientific Publishing Co. Pte. Ltd.
5 Toh Tuck Link, Singapore 596224
USA office: 27 Warren Street, Suite 401-402, Hackensack, NJ 07601
UK office: 57 Shelton Street, Covent Garden, London WC2H 9HE

British Library Cataloguing-in-Publication Data
A catalogue record for this book is available from the British Library.

THE MASS GAP AND ITS APPLICATIONS

ISBN 978-981-4440-70-7

Printed in Singapore by World Scientific Printers.

We dedicate this book

to our parents, wives, and children
who may read, but will not understand a word of it,
anyway we love them...

Preface

We should always keep in mind: Nature is tricky! This saying is multiply true for the *Standard Model* of particle physics. The properties of the strongly interacting matter, the neutrinos or the Higgs particle still remain unsolved parts of a great puzzle in this consistent, well-built, unified theory. Furthermore, recently it has become clear that such an important thing as the *vacuum* is one of the most complex objects in Nature.

In this book we focus on the most up-to-date theory of the strong interaction: Quantum Chromodynamics (QCD), which has been developed during the last decades of the 20th century. In high-energy collisions QCD has been proved by experimental tests, using perturbative treatments because of its *asymptotic freedom* behavior. It is the flip of Nature that these perturbative procedures become problematic if one tries to apply them to low-energy QCD — since the running coupling constant goes beyond 1. Due to *color confinement* the description of the well-known bound states like mesons, baryons (e.g. proton or neutron), and the vacuum are not yet available from first principles — the ultimate goal of any fundamental theory.

While we were finishing this book, the European Laboratory for Particle and Nuclear Physics (CERN, Geneva, Switzerland) made a statement on a Higgs-like particle. The missing puzzle of the Standard Model has been recognized at almost 5σ. This is a historical event, since by this discovery, the most wanted 'God's particle' might complete the best field theory we have for particle physics. On the other hand, finding the Higgs will not solve the problem of masses originating from the non-perturbative behavior of Quantum Chromodynamics.

Here, we present a new method to investigate non-perturbative QCD by the introduction of the 'mass gap into the theory, suggested first by

Arthur Jaffe and Edward Witten at the turn of this century. Our book is about this way of handling the mass problem in QCD. To explain the mass-spectrum of QCD we are going to show in this book that, for this aim one needs rather the mass scale parameter (the mass gap) than other massive particles. The mass gap is in principle responsible for the large-scale structure of the QCD ground state, and thus for its non-perturbative phenomena at low energies.

Simply speaking, the mass gap is the energy difference between the lowest order and the vacuum state in the Yang – Mills quantum field theory. In this book, we not only introduce and present the mass gap, but we give some applications and the outlook of the mass gap method. At some points we look ahead to the far future, thinking on the possible everyday use of the calculated results. In addition a detailed summary of references and problems are included.

We recommend this book for scientists and very advanced students who want to be familiar in more detail with non-perturbative effects and methods in QCD. This work starts there, where textbooks on Quantum Chromodynamics usually end, describing the strong interaction at high energies. However, we give a short introduction of QCD and its general features in Section 1.1. Readers are kindly advised to have a knowledge of the basics of the Quantum Chromodynamics. Readers, especially students, who would like to deal with this book in a more detailed way are suggested to study some of the articles and textbooks referred within the introductory section. Let us remark that problems were added to this book at the end of each chapter in order to help deepen the knowledge and techniques of the mass gap. We encourage students to work them out as a general feedback of the understanding.

Finally, the authors wish you an interesting journey to the interesting world of 'The Mass Gap and its Applications' on the following pages!

Vakhtang Gogokhia
Gergely Gábor Barnaföldi
Wigner Research Centre for Physics
Hungarian Academy of Science
Budapest, Hungary, 2012

Acknowledgments

The birth of this book would not have been possible without the ignition and permanent motivation of Júlia Nyíri and the Publishers. Later, once we started, it looked too far from the reality, but during the last year we started to see the light at the end of the tunnel. It is our great pleasure to thank all these for their support, including Julika's work on checking the text.

Both of us are working at the Department for Theoretical Physics of the Institute for Particle and Nuclear Physics, Wigner Research Center for Physics of the Hungarian Academy of the Sciences (before 2012 with name KFKI RMKI of the HAS). This stable and pleasant environment including our colleagues's help provides a good background for our work. One of us (V.G.) also acknowledges the support of the members of the A. Razmadze Mathematical Institute of I. Javakhishvili Tbilisi State University.

While we were working on the manuscript of the book, we were supported by Hungarian National Fund (OTKA) grants NK778816, NK106119, H07-C 74164, PD63596, and K104260. Author G.G.B. was also partially supported by NIH TET_10-1_2011-0061 and ZA-15/2009 grants and the János Bolyai Research Scholarship of the Hungarian Academy of Sciences.

Authors would like to thank their co-authors in some earlier papers: Gyula Kluge, Badri Magradze, Miklós Prisznyák, Tadakatsu Sakai, Hiroshi Toki, and Mátyás Vasúth. Furthermore, we are grateful to many people for useful remarks, suggestions, discussions, and help: Tamás Sándor Biró, Laszó Pál Csernai, Tamás Csörgő, Merab Eliashvili, George Fai, Vladimir Gribov, Péter Hraskó, Anzor Khelashvili, Ivan Kiguradze, Vladimir and Alexei Kouzushins, Alexandr Kvinikhidze, Károly Ladányi, Árpád Lukács, Béla Lukács, Péter Lévai, Nino Partsvania, Sona Pochybová, György Pócsik, Roman Shakarashvili, Avtandil Shurgaia, Vladimir

Skokov, Károly Szegő, Katalin Szép, Fridon Todua, Kálmán Tóth, Péter Ván, and József Zimányi.

Last, but not least, the authors would like to thank their parents and families for the patience and continuous support from the very beginning: Shota Gogokhia, Nina Mgeladze, Katalin Katona, Gyuri Gogohija, and Flóra Mészáros (V.G.); Henriett Éva Hirdi, Ágnes Nikolett Farkas, and Vivien Nóra Barnaföldi (G.G.B.).

Vakhtang Gogokhia
Gergely Gábor Barnaföldi
Wigner Research Centre for Physics
Hungarian Academy of Science
Budapest, Hungary, 2012

Contents

PART I
Theory of the Mass Gap

Chapter 1

Quantum Chromodynamics and the Mass Gap

1.1 Quantum Chromodynamics

Quantum Chromodynamics (QCD) [1–6] is widely accepted as a realistic, dynamical quantum gauge theory of strong interactions not only at the fundamental (microscopic) quark–gluon level, but at the hadronic (macroscopic) level as well. This means that, in principle, it should describe the properties of the observed hadrons in terms of never experimentally seen quarks and gluons, i.e., to describe the hadronic world from first principles — the ultimate goal of any fundamental dynamical theory. However, this is a formidable task because of the color confinement phenomenon, whose dynamical mechanism is not understood yet, and therefore the confinement problem remains unsolved up to the present day. It prevents colored quarks and gluons from being experimentally detected as asymptotic states, which are colorless (color-singlet) by definition, i.e., color is permanently confined, being thus absolute [2]. At present, there are no doubts left that color confinement as well as other dynamical effects, such as spontaneous/dynamical breakdown of chiral symmetry, bound-state problems, etc., are inaccessible to perturbative techniques, and thus they are very essential non-perturbative effects. In turn, this means that for their investigation, non-perturbative solutions, methods and approaches need to be found, developed and used. This is especially necessary taking into account that the above-mentioned non-perturbative effects are low-energy/momentum (large distances) phenomena and, as it is well known, the perturbative methods, in general, fail to investigate them.

The Lagrangian density, which describes the properties, symmetries and interactions between fundamental constituents (gluons and quarks) of QCD, can be given in the following simplified form based on,

$$L(x) = i\bar{q}^A_\alpha D_{\alpha\beta} q^A_\beta - \frac{1}{4} F^a_{\mu\nu} F^{a\mu\nu} \quad , \tag{1.1.1}$$

where space-time indices are $\mu, \nu = 0, 1, 2, 3$, and the color indices run as follows: $\alpha, \beta = 1, 2, 3$, while $a = 1, 2...8$. The number of different quarks, the so-called flavor number, is N_f and thus $A = 1, 2...N_f$. Let us also point out that a paper which provides an excellent guide with brief comments to the literature on QCD can be found in [7].

The gluon field strength tensor is given by

$$F^a_{\mu\nu} \equiv F^a_{\mu\nu}(x) = \partial_\mu A^a_\nu - \partial_\nu A^a_\mu + g f_{abc} A^b_\mu A^c_\nu \quad , \tag{1.1.2}$$

while the covariant derivative is defined as:

$$D_{\alpha\beta} q^A_\beta \equiv (\delta_{\alpha\beta}\partial_\mu - ig\frac{1}{2}\lambda^a_{\mu\nu} A^a_\mu)\gamma_\mu q^A_\beta \ . \tag{1.1.3}$$

Note that both quantities depend on the same coupling constant g, i.e., the Lagrangian of QCD has a single universal dimensionless coupling constant for all types of the interactions between the fundamental constituents. It is worth reminding here that the QCD fine-structure coupling constant $\alpha_s = g^2/4\pi$ (calculated at any scale) is much bigger numerically than its Quantum Electrodynamics (QED) counterpart. This restricts the application of the perturbation theory methods to QCD apart from in the limit of high energies due to asymptotic freedom phenomenon in this theory [8–10].

The λ^as generators are $SU(3)$ matrices, which obey the commutator relation below,

$$[\lambda^a, \lambda^b] = 2if^{abc}\lambda^c \tag{1.1.4}$$

with f^{abc} the structure constants of $SU(3)$ color gauge group.

One can verify that the Lagrangian (1.1.1) is invariant under local gauge transformation of the form

$$\begin{aligned} q^A &\rightarrow U(x)q^A(x), \\ A_\mu(x) = A^a_\mu(x)\lambda^a/2 &\rightarrow U(x)A_\mu(x)U^{-1}(x) + \frac{i}{g}U(x)\partial_\mu U^{-1}(x). \end{aligned} \tag{1.1.5}$$

Here the local $SU(3)$ color gauge transformation $U(x) = \exp(i\Theta^a(x)\lambda^a/2)$ is a function of space-time dependent parameters $\Theta^a(x)$. Thus equation (1.1.1) is the minimal locally gauge invariant Lagrangian density implied by this $SU(3)$ color symmetry.

The Lagrangian given by equation (1.1.1) is invariant under the larger chiral group

$$SU(N_f) \times SU(N_f) \times U_B(1) \times U_A(1) \tag{1.1.6}$$

with $q_{R,L} = \frac{1}{2}(1 \pm \gamma_5)q$ being the right and left hand-side components of the quark fields in the fundamental representation. $U_B(1)$ describes the baryon number conservation and $U_A(1)$ describes the axial-baryon number conservation, which is not wanted, since it is not observed. The chiral $SU(N_f) \times SU(N_f)$ flavor symmetry is broken in QCD by adding to equation (1.1.1) a quark mass term

$$L_q = \bar{q}^A_\alpha \delta^{AB} m_0 q^B_\alpha , \tag{1.1.7}$$

where m_0 is the so-called current quark mass, depending on flavor A. Let us underline that only this massive term is compatible with the $SU(3)$ color gauge symmetry of QCD. At the same time, the massive gluon term $m_g^2 A_\mu A_\mu$ explicitly violates it. By adding this term to Eq. (1.1.1), it is not invariant under local gauge transformations given by equations (1.1.5). This causes one of the important challenges of QCD.

1.2 The Jaffe – Witten theorem on the Mass Gap

Let us now bring the reader's attention to one of the important features of the Lagrangian of QCD briefly discussed above. It does not contain a mass scale parameter which could have a physical meaning. This is true even after the corresponding renormalization programme is performed. The current quarks are colored objects and that is why the hadron mass cannot directly depend on their masses: the color-singlet mass scale parameter is needed for this purpose. Precisely this important problem has been addressed at the beginning of this century by Arthur Jaffe and Edward Witten who have formulated one of the Millennium Prize Problems as follows [11]:

Yang – Mills existence and the Mass Gap: *Prove that for any compact simple gauge group G, quantum Yang – Mills theory on $\mathbb{R}^4$ exists and has a mass gap $\Delta > 0$.*

In the description of this theorem [11] they have explained why the mass gap is needed.

(i) It must have a 'mass gap'. Every excitation of the vacuum has energy at least Δ — to explain why the nuclear force is strong but short-range.
(ii) It must have 'quark confinement' — why the physical particles are $SU(3)$-invariant, i.e., color-independent.
(iii) It must have 'chiral symmetry breaking' — to account for the 'current algebra' theory of soft pions.

Summarizing, we need the mass gap which is responsible for the non-perturbative dynamics of QCD, and thus it determines the large-scale structure of the QCD ground state. Any hadron mass finally has to be expressed in terms of the renormalized mass gap itself, i.e., $M_h = \text{const}_h \times \Delta$, where h denotes any hadron, while const_h is the corresponding dimensionless constant. In other words, the hadron spectrum should depend on the mass gap. It is different from Λ_{QCD}, which is responsible for its non-trivial perturbative dynamics (scale violation, asymptotic freedom), and thus is due to the short-scale structure of QCD ground state. As we already know any mass term (for example, the gluon mass term), apart from the current quark masses, violates $SU(3)$ color gauge invariance/symmetry of QCD. However, in the next chapter we will show that the common belief (which comes from the perturbation theory) that the mass gap contradicts the above-mentioned color gauge symmetry is false. We will show that this fundamental symmetry is maintained/preserved at non-zero mass gap as well.

Problems

Problem 1.1. *Show explicitly that the QCD Lagrangian* (1.1.1) *is indeed invariant under local gauge transformation* (1.1.5).

Problem 1.2. *Show explicitly that the QCD Lagrangian* (1.1.1) *with the massive gluon term* $m_g^2 A_\mu A_\mu$ *included is indeed not invariant under local gauge transformation* (1.1.5).

Chapter 2

Color Gauge Invariance and the Origin of the Mass Gap

2.1 Introduction

As described above QCD is a $SU(3)$ color gauge invariant theory but:

(i) Due to color confinement, the gluon (unlike the photon) is not a physical state. Moreover, there is no physical amplitude to which the gluon self-energy (like the photon self-energy) may directly contribute.

(ii) In contrast to the conserved currents in QED, the color-conserved currents do not play any role in the extraction of physical information from the S-matrix elements for the corresponding physical processes and quantities in QCD. In other words, the conserved color currents do not contribute directly to the S-matrix elements describing this or that physical process/quantity. For this their color-singlet counterparts, which can even be partially conserved, are relevant. For example, an important physical QCD parameter such as the pion decay constant is given by the following S-matrix element:

$$\langle 0 \,|J^i_{5\mu}(0)|\, \pi^j(q) \rangle = iq_\mu F_\pi \delta^{ij}, \tag{2.1.1}$$

where $J^i_{5\mu}(0)$ is the axial-vector current, while $|\pi^j(q)\rangle$ describes the pion bound-state amplitude, and i, j are flavor indices.

(iii) In QCD (contrary to QED) there exists direct evidence/indication that transversality of the full gluon self-energy is violated. The Slavnov–Taylor identity for the full gluon propagator, as it is determined by the corresponding equation of motion, is also not satisfied. Indeed, there is no regularization scheme (preserving or not gauge invariance) in which the transversality condition and the Slavnov – Taylor identity could be satisfied unless the so-called constant skeleton tadpole term is put formally zero by hand everywhere.

Our main goal in this chapter is to show that the tadpole term is consistent with the color gauge invariance in QCD, i.e., it is maintained at the non-zero tadpole term as well.

However, some general remarks are in order. The surprising fact is that after more than forty years of QCD, we still don't know the interaction between quarks and gluons. To know it means that one knows the full gluon and quark propagators, the full quark–gluon and the pure gluon vertices. In the weak coupling limit or in the case of heavy quarks only this interaction is known. In the first case all the above-mentioned lower and higher Green's functions (propagators and vertices, respectively) become effectively free ones multiplied by the renormalization group perturbative logarithm improvements due to the asymptotic freedom of QCD [8–10]. In the case of heavy quarks all the Green's functions can be approximated by their free counterparts from the very beginning. In general, the Green's functions are essentially different from their free counterparts. They are substantially modified due to the response of the highly nontrivial structure of the true QCD vacuum. It is just this response which is taken into account by the full (dressed) propagators and vertices (it can be neglected in the weak coupling limit or for heavy quarks, as pointed out above). That is the main reason why they are still unknown, and the confinement and many other non-perturbative problems are not solved yet [12, 13].

In other words, it is not enough to know the Lagrangian of the theory (1.1.1). In QCD it is also necessary and important to know the true structure of its ground state (also there might be symmetries of the Lagrangian which do not coincide with symmetries of the vacuum). This knowledge can only come from the investigation of a general system of the dynamical equations of motion of QCD, the so-called Schwinger – Dyson system of equations [2, 14–19], which should be satisfied by all the Green's functions. Since this system of equations contains the full dynamical information on QCD, to solve it means to establish all the QCD Green's functions, and thus to establish the structure of the true QCD ground state as well. This system of equations is a highly non-linear, strongly coupled system of four-dimensional integral equations. In fact, it is an infinite chain of the relations between different propagators, vertices and scattering kernels. The kernels and scattering amplitudes of these integral equations are determined by an infinite series of the corresponding multi-loop skeleton diagrams. It is a general feature of non-linear systems that the number of solutions (if any) cannot be fixed *a priori*. Although this system of dynamical equations can be reproduced by expansion around the free field vacuum, the final equations make no reference to the vacuum of the perturbation theory. They are sufficiently general and should be treated beyond the perturbation theory [2].

These equations should be also complemented by the corresponding Slavnov – Taylor identities [2, 14, 15, 20–26], which, in general, relate lower and higher Green's functions to each other. These identities are consequences of the exact gauge invariance and therefore *"are exact constraints on any solution to QCD"* [2]. The low-energy/momentum region interesting for confinement is usually under the control of these identities. Precisely the Schwinger – Dyson system of dynamical equations, complemented by the Slavnov – Taylor identities, can serve as an adequate and effective tool for the non-perturbative approach to QCD.

2.2 The gluon Schwinger – Dyson equation

If there is no place for the mass scale parameter different from the current quark mass in the Lagrangian of QCD, then the only place where it may explicitly appear is the above-discussed system of dynamical equations of motion. This underlines the importance of the investigation of this system of equations and the corresponding identities for the understanding of the true dynamics in the QCD ground state. The propagation of gluons is one of the main dynamical effects there. The importance of the corresponding equation of motion is due to the fact that its solutions are supposed to reflect the quantum-dynamical structure of the QCD ground state. This equation is a highly non-linear one due to the self-interaction of massless gluon modes, so the number of its independent solutions is not fixed *a priori*, as emphasized above. From the very beginning they should be considered on equal footing. The color gauge structure of this equation will be investigated in this chapter in more detail.

For our purpose it is convenient to begin with the general description of Schwinger – Dyson equation for the full gluon propagator $D_{\mu\nu}(q)$. It can be written down as follows:

$$D_{\mu\nu}(q) = D^0_{\mu\nu}(q) + D^0_{\mu\rho}(q) i\Pi_{\rho\sigma}(q; D) D_{\sigma\nu}(q), \tag{2.2.1}$$

where

$$D^0_{\mu\nu}(q) = i\left[T_{\mu\nu}(q) + \xi L_{\mu\nu}(q)\right]\frac{1}{q^2} \tag{2.2.2}$$

is the free gluon propagator, and ξ is the gauge-fixing parameter. Also, here and everywhere below $T_{\mu\nu}(q) = \delta_{\mu\nu} - (q_\mu q_\nu/q^2) = \delta_{\mu\nu} - L_{\mu\nu}(q)$, as usual in Euclidean metrics (see remarks below). $\Pi_{\rho\sigma}(q; D)$ is the full gluon self-energy which depends on the full gluon propagator due to the non-abelian character of QCD. Evidently, we omit the color group indices, since for the gluon propagator (and hence for its self-energy) they factorize, for example $D^{ab}_{\mu\nu}(q) = D_{\mu\nu}(q)\delta^{ab}$.

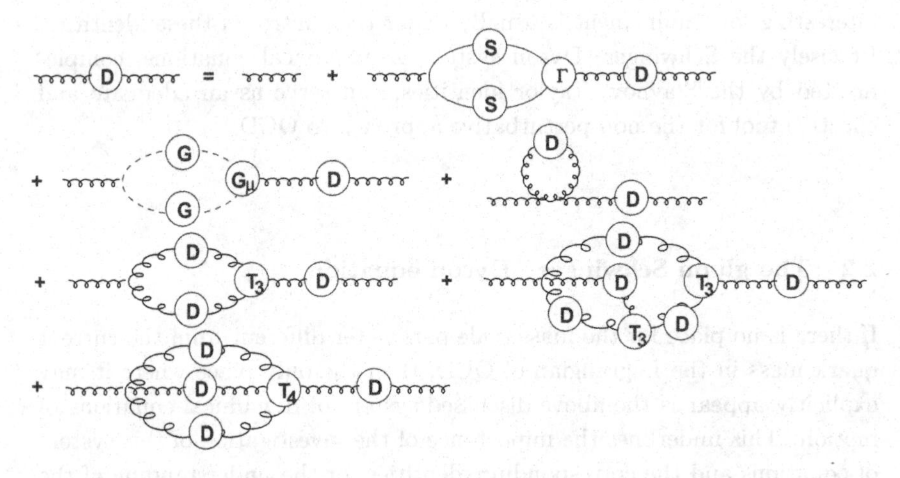

Fig. 2.2.1 The Schwinger – Dyson equation for the full gluon propagator.

Figure 2.2.1 shows that the full gluon self-energy, denoted by $\Pi_{\rho\sigma}(q; D)$, is the sum of a few terms,

$$\Pi_{\rho\sigma}(q; D) = \Pi^q_{\rho\sigma}(q) + \Pi^{gh}_{\rho\sigma}(q) + \Pi^t_{\rho\sigma}(D) + \Pi^{(1)}_{\rho\sigma}(q; D^2) + \Pi^{(2)}_{\rho\sigma}(q; D^4) + \Pi^{(2')}_{\rho\sigma}(q; D^3), \tag{2.2.3}$$

where $\Pi^q_{\rho\sigma}(q)$ describes the skeleton loop contribution due to the quark degrees of freedom (it is an analogue of the vacuum polarization tensor in QED). Let us also note that in this term the superscript 'q', which means quark, is not to be mixed up with the gluon momentum q. $\Pi^{gh}_{\rho\sigma}(q)$ describes the skeleton loop contribution associated with the ghost degrees of freedom. Since neither of the skeleton loop integrals depends on the full gluon propagator D, they represent the linear contribution to the gluon Schwinger–Dyson equation. $\Pi^{t}_{\rho\sigma}(D)$ is the so-called constant skeleton tadpole term. $\Pi^{(1)}_{\rho\sigma}(q; D^2)$ represents the skeleton loop contribution, which contains the triple gluon vertices only. $\Pi^{(2)}_{\rho\sigma}(q; D^4)$ and $\Pi^{(2')}_{\rho\sigma}(q; D^3)$ describe topologically independent skeleton two-loop contributions, which combine the triple and quartic gluon vertices. All these quantities are given by the corresponding skeleton loop diagrams in Fig. 2.2.1. The last four terms explicitly contain the full gluon propagators in the corresponding powers symbolically shown above. They thus form the non-linear part of the gluon Schwinger–Dyson equation. The analytical expressions for the corresponding skeleton loop integrals [27] (in which the symmetry coefficients and signs have been included, for convenience) are of no importance here, since we are not going to introduce into them any truncations/approximations/assumptions or choose some special gauge. Let us note in advance that here and below the signature is Euclidean, since it implies $q_i \to 0$ when $q^2 \to 0$ and *vice versa*.

All the quantities which contribute to the full gluon self-energy equation (2.2.3) are tensors, having the dimensions of mass squared. All these skeleton loop integrals are therefore quadratically divergent in perturbation theory, and so they are assumed to be regularized. However, this is not the whole story yet. Contrary to QED, QCD being a non-abelian gauge theory can suffer from infrared singularities in the $q^2 \to 0$ limit due to the self-interaction of massless gluon modes. Thus all the possible subtractions at zero may be dangerous [2]. That is why in all the quantities below the dependence on the finite (slightly different from zero) dimensionless subtraction point α is to be understood. In other words, all the subtractions at zero and the Taylor expansions around zero should be understood as the subtractions at α and the Taylor expansions near α, where they are justified to be used. From a technical point of view, however, it is convenient to put formally $\alpha = 0$ in all the expressions and derivations below, and to restore the explicit dependence on non-zero α in all the quantities only at the final stage. At the same time, in all the quantities where the dependence on λ (which is the dimensionless ultraviolet regulating parameter) and α is not shown explicitly, nevertheless, it should also be assumed. For example, $\Pi_{\rho\sigma}(q; D) \equiv \Pi_{\rho\sigma}(q; D, \lambda, \alpha)$ and similarly for all other quantities. This means that all the expressions are regularized (they become finite), and thus a mathematical meaning is assigned to all of them. For our purpose, in principle, it is not important how λ and α have been introduced. They should be removed at the final stage only as a result of the self-consistent renormalization program as well as all other un-physical parameters like the gauge-fixing, etc.

2.3 Transversality of the full gluon self-energy

Contracting the full gluon self-energy (2.2.3) with q_ρ, it can be reduced to the three independent transversal conditions, namely

$$q_\rho \Pi_{\rho\sigma}(q; D) = q_\rho \Pi^q_{\rho\sigma}(q) + q_\rho \Pi^g_{\rho\sigma}(q; D) + q_\rho \Pi^t_{\rho\sigma}(D), \tag{2.3.1}$$

where the gluon contribution, $\Pi^g_{\rho\sigma}(q; D)$ is defined as follows:

$$\Pi^g_{\rho\sigma}(q; D) = \Pi^{gh}_{\rho\sigma}(q) + \Pi^{(1)}_{\rho\sigma}(q; D^2) + \Pi^{(2)}_{\rho\sigma}(q; D^4) + \Pi^{(2')}_{\rho\sigma}(q; D^3). \tag{2.3.2}$$

The quark contribution

Quite similarly to the current conservation in QED [28], the color current conservation condition implies

$$q_\rho \Pi^q_{\rho\sigma}(q) = 0. \tag{2.3.3}$$

For further purpose it is instructive to introduce the subtracted quark contribution to the full gluon self-energy as follows:

$$\Pi^{q(s)}_{\rho\sigma}(q) = \Pi^q_{\rho\sigma}(q) - \Pi^q_{\rho\sigma}(0) = \Pi^q_{\rho\sigma}(q) - \delta_{\rho\sigma}\Delta^2_q, \tag{2.3.4}$$

where $\Pi^q_{\rho\sigma}(0) = \delta_{\rho\sigma}\Delta^2_q$ and Δ^2_q is the quadratically divergent, but already regularized constant, as underlined above (it is nothing but the corresponding skeleton loop integral at $q = 0$). The general decomposition of the quark part and its subtracted counterpart into the independent tensor structures are

$$\begin{aligned} \Pi^q_{\rho\sigma}(q) &= T_{\rho\sigma}(q) q^2 \Pi^q(q^2) + q_\rho q_\sigma \tilde{\Pi}^q(q^2), \\ \Pi^{q(s)}_{\rho\sigma}(q) &= T_{\rho\sigma}(q) q^2 \Pi^{q(s)}(q^2) + q_\rho q_\sigma \tilde{\Pi}^{q(s)}(q^2). \end{aligned} \tag{2.3.5}$$

Here and everywhere below all the invariant functions are dimensionless ones of their argument q^2. In addition both invariant functions $\Pi^{q(s)}(q^2)$ and $\tilde{\Pi}^{q(s)}(q^2)$ cannot have power-type singularities (or, equivalently, pole-type ones) at small q^2, since $\Pi^{q(s)}_{\rho\sigma}(0) = 0$ by definition in equation (2.3.4): otherwise they remain arbitrary. On account of the subtraction (2.3.4), one obtains

$$\Pi^q(q^2) = \Pi^{q(s)}(q^2) + \frac{\Delta_q^2}{q^2},$$
$$\tilde{\Pi}^q(q^2) = \tilde{\Pi}^{q(s)}(q^2) + \frac{\Delta_q^2}{q^2}, \tag{2.3.6}$$

then the quark contribution to the full gluon self-energy becomes

$$\Pi^q_{\rho\sigma}(q) = T_{\rho\sigma}(q)\left[q^2\Pi^{q(s)}(q^2) + \Delta_q^2\right] + L_{\rho\sigma}\left[q^2\tilde{\Pi}^{q(s)}(q^2) + \Delta_q^2\right]. \tag{2.3.7}$$

However, from the color current conservation condition (2.3.3) it follows that

$$\tilde{\Pi}^{q(s)}(q^2) = -\frac{\Delta_q^2}{q^2}, \tag{2.3.8}$$

which is impossible since $\tilde{\Pi}^{q(s)}(q^2)$ cannot have power-type singularities at small q^2, as underlined above. The only solution for the previous relation is to disregard Δ_q^2 on a general ground, i.e., put formally zero, and hence $\tilde{\Pi}^q(q^2) = \tilde{\Pi}^{q(s)}(q^2) = 0$ as well. So, one has

$$\Delta_q^2 = 0, \tag{2.3.9}$$

and the subtracted quark part $\Pi^{q(s)}_{\rho\sigma}(q)$ becomes also transversal $q_\rho\Pi^{q(s)}_{\rho\sigma}(q) = 0$, and $\Pi^q(q^2) = \Pi^{q(s)}(q^2)$, which yields $\Pi^q_{\rho\sigma}(q) = \Pi^{q(s)}_{\rho\sigma}(q)$. Hence the general decomposition of the quark part is

$$\Pi^q_{\rho\sigma}(q) = T_{\rho\sigma}(q)q^2\Pi^{q(s)}(q^2) \tag{2.3.10}$$

with the invariant dimensionless function $\Pi^{q(s)}(q^2)$ having no pole-type singularities in the $q^2 \to 0$ limit, i.e., in fact, it is a regular function in this limit. It is also free of the quadratic divergences, which are incorporated into Δ_q^2, i.e., it does not depend on it because of the condition (2.3.9). This describes a general situation when the initial transversal condition (2.3.3) for $\Pi^q_{\rho\sigma}(q)$ transforms the quadratic divergence of the corresponding loop integral(s) to a logarithmic divergence at large q^2. They may still be present in the invariant function $\Pi^q(q^2)$, and hence in the invariant function $\Pi^{q(s)}(q^2)$. This is in complete analogy with QED (see again [28] and Appendix 2.A), since there only electron–positron skeleton loop (the vacuum polarization tensor) contributes to the full photon self-energy.

Concluding, let us note the the quark constant has to be always disregarded, see equation (2.3.9), since due to the color current conservation condition the transversality condition (2.3.3) always satisfied. The skeleton quark loop contribution to the full gluon self-energy can always be made transversal independently from the gluon parts.

The gluon contribution

It is well known that in QCD the gluon part is also transversal, namely

$$q_\rho \Pi^g(q;D) = q_\rho \Big[\Pi^{gh}_{\rho\sigma}(q) + \Pi^{(1)}_{\rho\sigma}(q;D^2) + \Pi^{(2)}_{\rho\sigma}(q;D^4) + \Pi^{(2')}_{\rho\sigma}(q;D^3)\Big] = 0. \tag{2.3.11}$$

It should be noted that none of these quantities can satisfy this transversal condition separately from each other, i.e, similarly to the relation (2.3.3). The role of ghost degrees of freedom is to cancel the un-physical (longitudinal) component of the full gluon propagator. Therefore the transversal condition (2.3.11) is important for ghosts to fulfill their role, and thus to maintain unitarity of the S-matrix in QCD. More precisely, just the ghost contribution $\Pi^{gh}_{\rho\sigma}(q)$ makes $\Pi^g(q^2;D)$ transversal. In absolutely the same way, as we have investigated the quark case in the previous subsection, it is possible to develop the general formalism for the gluon part as well. So let us begin with the corresponding subtraction

$$\Pi^{g(s)}_{\rho\sigma}(q;D) = \Pi^g_{\rho\sigma}(q;D) - \Pi^g_{\rho\sigma}(0;D) = \Pi^g_{\rho\sigma}(q;D) - \delta_{\rho\sigma}\Delta^2_g(D), \tag{2.3.12}$$

where

$$\Delta^2_g(D) = \Pi^g(0;D) = \sum_a \Pi_a(0;D) = \sum_a \Delta^2_a(D), \tag{2.3.13}$$

and the index 'a' runs as follows: $a = gh, (1), (2), (2')$, while Δ^2_{gh} does not depend on D. In this relation all the quadratically divergent constants $\Delta^2_a(D)$, having the dimensions of mass squared, are given by the corresponding regularized skeleton loop integrals at $q^2 = 0$ that appear in equations (2.3.11), and (2.2.3).

The general decomposition of the gluon part and its subtracted counterpart into the independent tensor structures are

$$\Pi^g_{\rho\sigma}(q;D) = T_{\rho\sigma}(q)q^2\Pi^g(q^2;D) + q_\rho q_\sigma \tilde{\Pi}^g(q^2;D),$$
$$\Pi^{g(s)}_{\rho\sigma}(q;D) = T_{\rho\sigma}(q)q^2\Pi^{g(s)}(q^2;D) + q_\rho q_\sigma \tilde{\Pi}^{g(s)}(q^2;D). \tag{2.3.14}$$

Here and everywhere below all the invariant functions are dimensionless ones of their argument q^2. As we already know, in addition both invariant functions $\Pi^{g(s)}(q^2;D)$ and $\tilde{\Pi}^{g(s)}(q^2;D)$ cannot have power-type singularities (or, equivalently, pole-type ones) at small q^2, since $\Pi^{g(s)}_{\rho\sigma}(0;D) = 0$ by definition in equation (2.3.12): otherwise they remain arbitrary. On account of the subtraction (2.3.12), one obtains

$$\Pi^g(q^2;D) = \Pi^{g(s)}(q^2;D) + \frac{\Delta_g^2(D)}{q^2},$$
$$\tilde{\Pi}^g(q^2;D) = \tilde{\Pi}^{g(s)}(q^2;D) + \frac{\Delta_g^2(D)}{q^2}, \tag{2.3.15}$$

then the gluon contribution to the full gluon self-energy becomes

$$\Pi^g_{\rho\sigma}(q;D)$$
$$= T_{\rho\sigma}(q)\left[q^2\Pi^{g(s)}(q^2;D) + \Delta_g^2(D)\right] + L_{\rho\sigma}\left[q^2\tilde{\Pi}^{g(s)}(q^2;D) + \Delta_g^2(D)\right]. \tag{2.3.16}$$

The transversality condition (2.3.11) implies

$$\tilde{\Pi}^{g(s)}(q^2;D) = -\frac{\Delta_g^2(D)}{q^2}, \tag{2.3.17}$$

as it follows from the previous expression. However, this is impossible, since the invariant function $\tilde{\Pi}^{g(s)}(q^2;D)$ cannot have pole-type singularities as underlined above. As in the quark case, the only solution to this condition is to disregard $\Delta_g^2(D)$ on a general ground, i.e., put formally zero in all relations, equations, which means that $\tilde{\Pi}^g(q^2;D) = \tilde{\Pi}^{g(s)}(q^2;D) = 0$ as well. Doing so

$$\Delta_g^2(D) = 0, \tag{2.3.18}$$

the subtracted gluon part $\Pi^{g(s)}_{\rho\sigma}(q)$ becomes also transversal $q_\rho \Pi^{g(s)}_{\rho\sigma}(q) = 0$, and $\Pi^g(q^2; D) = \Pi^{g(s)}(q^2; D)$, which yields $\Pi^g_{\rho\sigma}(q; D) = \Pi^{g(s)}_{\rho\sigma}(q; D)$.

Thus, in fact, the general decomposition of the gluon part is

$$\Pi^g_{\rho\sigma}(q; D) = T_{\rho\sigma}(q) q^2 \Pi^{g(s)}(q^2; D), \tag{2.3.19}$$

and the invariant dimensionless function $\Pi^{g(s)}(q^2; D)$ has the same properties as the quark invariant function $\Pi^{q(s)}(q^2)$, i.e., despite depending on D it is a regular function of its argument q^2.

Concluding, for the explicit demonstration of how the ghosts guarantee the transversality condition (2.3.11) in lower order of the perturbation theory see, for example [3, 5]. However, this should be true in every order of the perturbation theory, thus going beyond it, in agreement with (2.3.11) where the skeleton loop integrals are present.

The tadpole term contribution

The transversality of the full gluon self-energy is always violated by the tadpole term, since

$$q_\rho \Pi_{\rho\sigma}(q; D) = q_\rho \Pi^t_{\rho\sigma}(D) = q_\sigma \Delta^2_t(D) \neq 0, \tag{2.3.20}$$

where $\Pi^t_{\rho\sigma}(D) = \delta_{\rho\sigma}\Delta^2_t(D)$ and $\Delta^2_t(D)$ is the corresponding quadratically divergent but regularized constant (the corresponding skeleton loop integral in equation (2.2.3)). It is important to understand that the tadpole term violates the transversality condition independently from all other contributions to the full gluon self-energy. To show this explicitly let us continue with the general tensor decomposition of the full gluon self-energy as follows:

$$\Pi_{\rho\sigma}(q; D) = T_{\rho\sigma}(q) q^2 \Pi_f(q^2; D) + q_\rho q_\sigma \tilde{\Pi}_f(q^2; D), \tag{2.3.21}$$

where the subscript 'f' shows that these invariant functions enter the full gluon self-energy. Contracting it with q_ρ, one obtains

$$\tilde{\Pi}_f(q^2; D) = \frac{\Delta^2_t(D)}{q^2}, \tag{2.3.22}$$

on account of the relation (2.3.20). So the full gluon self-energy becomes

$$\Pi_{\rho\sigma}(q; D) = T_{\rho\sigma}(q) q^2 \Pi_f(q^2; D) + L_{\rho\sigma}(q)\Delta^2_t(D), \tag{2.3.23}$$

and it is not transversal indeed because of the tadpole term. The tadpole term contributes to the invariant function $\Pi_f(q^2; D)$ as well. To show this let us remind that from the relation (2.3.1) it follows that

$$\Pi_{\rho\sigma}(q; D) = \Pi^q_{\rho\sigma}(q) + \Pi^g_{\rho\sigma}(q; D) + \delta_{\rho\sigma}\Delta^2_t(D). \tag{2.3.24}$$

Equating this expression to the previous one, on account of equations (2.3.10) and (2.3.19), and doing some algebra, one arrives at

$$q^2\Pi_f(q^2;D) = q^2\left[\Pi^{q(s)}(q^2) + \Pi^{g(s)}(q^2;D)\right] + \Delta_t^2(D). \qquad (2.3.25)$$

Thus the full gluon self-energy (2.3.23) finally becomes

$$\Pi_{\rho\sigma}(q;D) = T_{\rho\sigma}(q)\left[q^2\Pi(q^2;D) + \Delta_t^2(D)\right] + L_{\rho\sigma}(q)\Delta_t^2(D), \qquad (2.3.26)$$

where in the first term

$$\Pi(q^2;D) \equiv \Pi^s(q^2;D) = \Pi^{q(s)}(q^2) + \Pi^{g(s)}(q^2;D)\ , \qquad (2.3.27)$$

so it remains a regular function of q^2.

Concluding, in the explicit presence of the tadpole term the ghosts cannot make the full gluon propagator transversal. To make it transversal one has to put it zero by hand everywhere. We are not going to do this from the very beginning, since our aim will be to maintain it in the transversal part of the full gluon self-energy (2.3.26). In this connection, let us remind that the constants Δ_q^2 and $\Delta_g^2(D)$ have been disregarded on a general ground, and not put to zero by hand, since the corresponding invariant functions could not have pole-type singularities along which these constants come. Let us also note in what follows we will omit the subscript 't' in the notation of the tadpole term, we put $\Delta_t^2(D) \equiv \Delta^2(D)$ and in what follows we call it the mass scale parameter.

2.4 Slavnov – Taylor identity for the full gluon propagator

In order to calculate the physical observables in QCD from first principles, we need the full gluon propagator rather than the full gluon self-energy. The basic relation to which the full gluon propagator should satisfy is the corresponding Slavnov – Taylor identity

$$q_\mu q_\nu D_{\mu\nu}(q) = i\xi. \qquad (2.4.1)$$

It is a consequence of the color gauge invariance/symmetry of QCD, and therefore "is an exact constraint on any solution to QCD" as emphasized in [2]. This is true for any other Slavnov – Taylor identity. Being a result of this exact symmetry, it is a general one, and it is important for the renormalization of QCD. If some equation, relation or the regularization scheme, etc. do not satisfy it automatically, i.e., without any additional conditions, then they should be modified and not this identity. In other words, all the relations, equations, regularization schemes, etc. should be adjusted to it and not *vice versa*. It implies that general tensor decomposition of the full gluon propagator is

$$D_{\mu\nu}(q) = i\left[T_{\mu\nu}(q)d(q^2) + \xi L_{\mu\nu}(q)\right]\frac{1}{q^2}, \qquad (2.4.2)$$

where the invariant function $d(q^2)$ is the corresponding Lorentz structure of the full gluon propagator (sometimes we will call it the full effective (running) charge, for simplicity). Let us emphasize once more that these basic relations are to be satisfied in any case, for example, independently of the fact whether the tadpole term itself or any other mass scale parameter is put formally zero or not.

Substituting equation (2.3.26) into the initial gluon Schwinger–Dyson equation (2.2.1), it can be equivalently re-written as follows:

$$\begin{aligned} D_{\mu\nu}(q) = D^0_{\mu\nu}(q) &+ D^0_{\mu\rho}(q)iT_{\rho\sigma}(q)\left[q^2\Pi(q^2;D) + \Delta^2(D)\right]D_{\sigma\nu}(q) \\ &+ D^0_{\mu\rho}(q)iL_{\rho\sigma}(q)\Delta^2(D)D_{\sigma\nu}(q). \end{aligned} \qquad (2.4.3)$$

Contracting this equation with q_μ and q_ν, one finally arrives at

$$q_\mu q_\nu D_{\mu\nu}(q) = i\xi\left[1 - \xi\frac{\Delta^2(D)}{q^2}\right], \qquad (2.4.4)$$

so the Slavnov–Taylor identity (2.4.1) is not automatically satisfied. In order to get from this relation the Slavnov–Taylor identity, one needs to put

$$\Delta^2(D) = 0 \qquad (2.4.5)$$

by hand. Thus the one way to satisfy the Slavnov–Taylor identity and thus to maintain the color gauge structure of QCD is to discard the mass scale parameter $\Delta^2(D)$ from the very beginning, i.e., put it formally zero in all the equations, relations, etc. — everywhere. What is the most unpleasant thing that the mass scale parameter has to be neglected in the transversal part of the Schwinger–Dyson equation (2.4.3) as well. First of all it defines the dynamics of the initial full gluon propagator, and secondly its presence in this part does not violate the Slavnov–Taylor identity.

However, in the formal $\Delta^2(D) = 0$ limit the initial gluon Schwinger–Dyson equation (2.4.3) is modified to

$$D^{PT}_{\mu\nu}(q) = D^0_{\mu\nu}(q) + D^0_{\mu\rho}(q)iT_{\rho\sigma}(q)q^2\Pi(q^2;D^{PT})D^{PT}_{\sigma\nu}(q), \qquad (2.4.6)$$

and the corresponding Lorentz structure which appears in equation (2.4.2)

$$D^{PT}_{\mu\nu}(q) = i\left[T_{\mu\nu}(q)d^{PT}(q^2) + \xi L_{\mu\nu}(q)\right]\frac{1}{q^2} \qquad (2.4.7)$$

becomes

$$d^{PT}(q^2) = \frac{1}{1 + \Pi(q^2; D^{PT})}. \tag{2.4.8}$$

It is easy to see that the modified gluon Schwinger–Dyson equation (2.4.6) now automatically satisfies the Slavnov–Taylor identity (2.4.1). The transversality condition (2.3.20) for the full gluon self-energy (2.3.26) is satisfied as well, namely

$$\Pi_{\rho\sigma}(q; D^{PT}) = T_{\rho\sigma}(q) q^2 \Pi(q^2; D^{PT}), \quad q_\rho \Pi_{\rho\sigma}(q; D^{PT}) = 0. \tag{2.4.9}$$

In the formal $\Delta^2(D) = 0$ limit we denote $D_{\mu\nu}(q)$ and $d(q^2)$ as $D^{PT}_{\mu\nu}(q)$ and $d^{PT}(q^2)$, respectively, where superscript 'PT' is for perturbation theory.

Concluding, in the formal $\Delta^2(D) = 0$ limit there will be no problems for ghosts to accomplish their role, namely to cancel the longitudinal component in the gluon propagator which satisfies equation (2.4.6). We have reminded some important aspects of the color gauge structure of QCD, but without any use of the perturbation theory. Also, in obtaining these results no regularization scheme (preserving or not gauge invariance) has been used. No special gauge choice has been made either.

2.5 The general structure of the full gluon propagator

The formal $\Delta^2(D) = 0$ limit is one way how to preserve the color gauge invariance in QCD. Then a natural question arises: why does the mass scale parameter $\Delta^2(D)$ (which is nothing but the tadpole term) exist in this theory at all? There is no doubt that the color gauge invariance of QCD should be maintained with non-zero tadpole term as well, since it is explicitly present in the full gluon self-energy, and hence in the full gluon propagator. However, by keeping it 'alive', two important problems arise.

(i) The first and most important problem is how to maintain $\Delta^2(D)$ in the transversal part of the initial Schwinger–Dyson equation (2.4.3), *and at the same time, to remove it from its longitudinal counterpart (which violates the Slavnov–Taylor identity) but without going to the formal $\Delta^2 = 0$ limit.*

The second problem is that even if the general mass scale parameter $\Delta^2(D)$ in the transversal part of the full gluon propagator (2.4.2) is retained, ghosts cannot cancel its longitudinal component. This means that the transversality of the full gluon self-energy will always be violated (see the general relation (2.3.20)). So there is no solution of this problem at the level of the gluon self-energy. However, we have previously noticed that, in principle, we need rather the relevant gluon propagator to be transversal at non-zero $\Delta^2(D)$ than the full gluon self-energy. Exactly how to avoid this difficulty (and thus to neutralize its negative consequences) for the relevant gluon propagator at non-zero $\Delta^2(D)$, i.e. to make it transversal without ghosts, will be explained and advocated below.

(ii) The second problem is how to make the full gluon propagator purely transversal when the mass scale parameter is explicitly present in its transversal component.

The spurious mechanism

To solve the first problem, i.e., to keep the general mass scale parameter 'alive', and, at the same time, to satisfy the Slavnov – Taylor identity (2.4.1), it is instructive to introduce a temporary dependence on $\Delta^2(D)$ in the free gluon propagator, thus making it an auxiliary (spurious) free gluon propagator. The original gluon Schwinger – Dyson equation (2.4.3) then should read

$$\begin{aligned} D_{\mu\nu}(q) &= D^0_{\mu\nu}(q;\Delta^2(D)) \\ &\quad + D^0_{\mu\rho}(q;\Delta^2(D)) iT_{\rho\sigma}(q)\left[q^2\Pi(q^2;D) + \Delta^2(D)\right] D_{\sigma\nu}(q) \\ &\quad + D^0_{\mu\rho}(q;\Delta^2(D)) iL_{\rho\sigma}(q)\Delta^2(D) D_{\sigma\nu}(q) \end{aligned} \tag{2.5.1}$$

with the spurious free gluon propagator as follows:

$$D^0_{\mu\nu}(q;\Delta^2(D)) = D^0_{\mu\nu}(q) + i\xi L_{\mu\nu}(q) d_0(q^2;\Delta^2(D))\frac{1}{q^2}. \tag{2.5.2}$$

The spurious free gluon propagator $D^0_{\mu\nu}(q;\Delta^2(D))$ can deviate from the standard free gluon propagator (2.2.2) only in its longitudinal component. The latter automatically satisfies the Slavnov – Taylor identity (2.4.1), while for the former this may not be true, indeed. Substituting this sum into the gluon Schwinger – Dyson (2.5.1), one obtains

$$\begin{aligned} D_{\mu\nu}(q) &= D^0_{\mu\nu}(q) + D^0_{\mu\rho}(q) iT_{\rho\sigma}(q)\left[q^2\Pi(q^2;D) + \Delta^2(D)\right] D_{\sigma\nu}(q) \\ &\quad + I_{\mu\nu}(q;\Delta^2(D)), \end{aligned} \tag{2.5.3}$$

where the final expression for $I_{\mu\nu}(q; \Delta^2(D))$ becomes

$$I_{\mu\nu}(q) = i\xi L_{\mu\nu}(q)\left[d_0(q^2; \Delta^2(D)) - \xi[1 + d_0(q^2)]\frac{\Delta^2(D)}{q^2}\right]\frac{1}{q^2}. \tag{2.5.4}$$

Just this term violates the Slavnov – Taylor identity (2.4.1) in the gluon Schwinger – Dyson equation (2.5.3), so it should be zero, which implies

$$d_0(q^2; \Delta^2(D)) = \xi\left[1 + d_0(q^2; \Delta^2(D))\right]\frac{\Delta^2(D)}{q^2}. \tag{2.5.5}$$

The explicit expression for $d_0(q^2; \Delta^2(D))$ is not important, since nothing in what follows will depend on it. Evidently, from now on we can completely forget about the spurious free gluon propagator. It played its role and retired from the scene.

Thus, the gluon Schwinger – Dyson equation. (2.5.3) finally becomes

$$D_{\mu\nu}(q) = D^0_{\mu\nu}(q) + D^0_{\mu\rho}(q)iT_{\rho\sigma}(q)\left[q^2\Pi(q^2; D) + \Delta^2(D)\right]D_{\sigma\nu}(q) \tag{2.5.6}$$

with the invariant function in equation (2.4.2) becoming

$$d(q^2) = \frac{1}{1 + \Pi(q^2; D) + (\Delta^2(D)/q^2)}. \tag{2.5.7}$$

The important observation is that now it is not required to put the mass scale parameter $\Delta^2(D)$ formally zero everywhere.

The spurious mechanism does not affect the dynamical context of the original gluon Schwinger – Dyson equation (2.4.3), which is determined by its transversal part equal to equation (2.5.7). It makes it possible to retain the mass scale parameter in the transversal part of the gluon Schwinger – Dyson equation, and, at the same time, to remove it from its longitudinal part, which violates the Slavnov – Taylor identity, but without going to the formal $\Delta^2(D) = 0$ limit everywhere. In this way, the modified gluon Schwinger – Dyson equation (2.5.6) satisfies automatically the Slavnov – Taylor identity. In other words, adding or subtracting longitudinal component to the initial Schwinger – Dyson equation in any way — in particular by the spurious method (see equation (2.5.1)) — leaves its 'physical' context unchanged. But we have shown, in the $\Delta^2(D) = 0$ limit it is drastically changed. That is why we consider the modified gluon Schwinger – Dyson equation (2.5.6) as more general than its original counterpart (2.4.3), which does not satisfy the Slavnov – Taylor identity.

Concluding, the relation (2.5.7) cannot be considered as a formal solution for the full gluon propagator in equation (2.4.2). The invariant function $d(q^2)$ depends on D via the mass scale term contribution $(\Delta^2(D)/q^2)$. The invariant function $\Pi(q^2; D)$ though formally depending on D, nevertheless is a regular function at small q^2, and it can be only logarithmically divergent at $q^2 \to \infty$: otherwise it remains arbitrary. Thus, in fact, $\Pi(q^2; D) \equiv \Pi(q^2; D^{PT})$, but we retain the previous notation until Section 3.2.

2.6 Non-perturbative vs. Perturbative QCD

In the previous Section it has been explicitly shown how the gluon Schwinger–Dyson equation should be modified in order to automatically satisfy the Slavnov–Taylor identity at non-zero mass scale parameter. It is instructive to collect our results here.

Non-perturbative QCD

The modified gluon Schwinger–Dyson equation (2.5.6) is

$$D_{\mu\nu}(q; \Delta^2(D)) = D^0_{\mu\nu}(q) + D^0_{\mu\rho}(q) i T_{\rho\sigma}(q) \left[q^2\Pi(q^2; D) + \Delta^2(D)\right] D_{\sigma\nu}(q; \Delta^2(D)), \tag{2.6.1}$$

while its general tensor decomposition is the standard one, namely

$$D_{\mu\nu}(q; \Delta^2(D)) = i \left[T_{\mu\nu}(q) d(q^2; \Delta^2(D)) + \xi L_{\mu\nu}(q)\right] \frac{1}{q^2}, \tag{2.6.2}$$

and its Lorentz structure is again given by equation (2.5.7), namely

$$d(q^2; \Delta^2(D)) = \frac{1}{1 + \Pi(q^2; D) + (\Delta^2(D)/q^2)}. \tag{2.6.3}$$

This system of equations forms the system of equations for the non-perturbative QCD, since we distinguish between the non- and the perturbative QCD by the explicit presence of the mass scale parameter, see discussion below — that is why we temporarily introduce the dependence on it in all the quantities above.

Perturbative QCD

The complete set of equations for the perturbative QCD is

$$D^{PT}_{\mu\nu}(q) = D^0_{\mu\nu}(q) + D^0_{\mu\rho}(q) i T_{\rho\sigma}(q) q^2 \Pi(q^2; D^{PT}) D^{PT}_{\sigma\nu}(q), \tag{2.6.4}$$

with

$$D_{\mu\nu}^{PT}(q) = i\left[T_{\mu\nu}(q)d^{PT}(q^2) + \xi L_{\mu\nu}(q)\right]\frac{1}{q^2}, \tag{2.6.5}$$

and its Lorentz structure is

$$d^{PT}(q^2) = \frac{1}{1+\Pi(q^2; D^{PT})}. \tag{2.6.6}$$

In both systems of equations the free gluon propagator is given by equation (2.2.2). The non-perturbative QCD system of equations has been obtained with the help of the spurious mechanism. It made it possible to keep the mass scale parameter $\Delta^2(D)$ alive, and, at the same time, to automatically satisfy the Slavnov – Taylor identity. The perturbative QCD system of equations has been obtained by putting the mass scale parameter formally zero everywhere. Evidently, the perturbative system of equations can be obtained from the non-perturbative system of equations in the formal $\Delta^2(D) = 0$ limit, since the dependence of the latter system of equations on the mass scale parameter $\Delta^2(D)$ is a regular one.

Due to asymptotic freedom in QCD the perturbative regime is realized at $q^2 \to \infty$. In this limit all the Green's functions are possible to approximate by their free perturbative counterparts (up to the corresponding perturbative logarithms). However, from the relation (2.6.3) it follows that in this limit the mass scale term contribution $\Delta^2(D)/q^2$ is only a next-to-next-to-leading order one. The leading order contribution is the invariant function $\Pi(q^2; D)$, which behaves like $\ln q^2$ in this limit, as mentioned above. The constant 1 is the next-to-leading order term in the $q^2 \to \infty$ limit. Such a special structure of the relation (2.6.3), namely the mass scale parameter enters it through the combination $\Delta^2(D)/q^2$ in its denominator only, explains immediately why the mass scale parameter $\Delta^2(D)$ is not important in the perturbation theory. From this structure it follows that the perturbative regime at $q^2 \to \infty$ is effectively equivalent to the formal $\Delta^2(D) = 0$ limit and *vice versa*. That is the reason why this limit can be called the perturbation theory limit. And that is why we denote $D_{\mu\nu}(q; \Delta^2 = 0) = D_{\mu\nu}(q; 0) \equiv D_{\mu\nu}^{PT}(q)$, and hence $d(q^2; \Delta^2 = 0) = d(q^2; 0) \equiv d^{PT}(q^2)$, etc., in accordance with the previous notations. However, sometimes it is useful to distinguish between the asymptotic suppression of the mass scale term contribution $\Delta^2(D)/q^2$ in the $q^2 \to \infty$ limit and the formal perturbation theory $\Delta^2(D) = 0$ limit — for simplicity, in what follows we call it the formal $\Delta^2(D) = 0$ limit.

Thus the formal $\Delta^2(D) = 0$ limit exists, and it is a regular one. As it follows from above, in this limit one recovers the perturbative QCD system of equations from the non-perturbative QCD one. So, we distinguish between the perturbative and the non-perturbative phases in QCD by the explicit presence of the mass scale parameter. Its aim is to be responsible for the non-perturbative QCD dynamics, since it dominates over the first two terms in the $q^2 \to 0$ limit in the solution (2.6.3). When it is put formally zero, then the perturbative phase survives only. Evidently, when such a scale is explicitly present then the QCD coupling constant plays no role in the non-perturbative QCD dynamics.

The mass scale parameter term does not survive in the perturbative $q^2 \to \infty$ regime, anyway. Then it is justified to simply drop it in the perturbation theory. It is worth emphasizing that this does not depend on how it has been regularized. However, as underlined above, any regularization scheme should be adjusted to the Slavnov–Taylor identity (2.4.1). In fact, in the most popular dimensional regularization method [29] it is prescribed to put $\Delta^2(D_0) = 0$ (see also [3, 5] and especially the corresponding discussion in [4]). So this method preserves the color gauge invariance in perturbative QCD from the very beginning.

2.7 The Mass Gap

The explicit presence of the constant skeleton tadpole term (which is nothing but the mass scale parameter) in the full gluon propagator is no coincidence. On the one hand, it does not contradict the color gauge invariance of four-dimensional QCD. As it has been explicitly shown so far in this chapter it is compatible with the Slavnov–Taylor identity. On the other hand, it will make it possible to introduce the mass gap so needed in the non-perturbative QCD in order to explain color confinement and other non-perturbative effects [11].

Let us note here that in two-dimensional QCD [30–32] the transversal condition for the full gluon self-energy is satisfied. This means that the tadpole term should be included from the very beginning. Otherwise, the ghosts will not be able to cancel the longitudinal component of the full gluon propagator [3]. However, this theory has already the scale parameter of dimension mass, which is the coupling constant. This once more emphasizes the special status of the tadpole term, and hence of the mass scale parameter, in four-dimensional QCD.

For further discussion it is convenient to re-write the relation (2.6.3) in the form of the corresponding transcendental equation, namely

$$d(q^2) = 1 - \left[\Pi(q^2; d) + \frac{\Delta^2(d)}{q^2} \right] d(q^2), \tag{2.7.1}$$

where instead of D we introduce an equivalent dependence on d, so that $\Delta^2(D) \equiv \Delta^2(d)$. The mass scale parameter $\Delta^2(d) \equiv \Delta^2(\lambda, \alpha, \xi, g^2; d)$, where g^2 is the dimensionless coupling constant squared, can be presented as follows:

$$\Delta^2(\lambda, \alpha, \xi, g^2; d) = \Delta^2 \times c(\lambda, \alpha, \xi, g^2; d), \tag{2.7.2}$$

where the mass squared

$$\Delta^2 \equiv \Delta^2(\lambda, \alpha; \xi, g^2), \tag{2.7.3}$$

will be called the mass gap. Contrary to the arbitrary dimensionless constant $c(\lambda, \alpha, \xi, g^2; d)$, it does not depend on d, but may, in general, depend on $\lambda, \alpha, \xi, g^2$, and so on. Thus at this stage it is only regularized as well as the mass scale parameter itself. Let us clarify that independence of the mass gap on d should be understood in the way that, in fact, it is the same for any d.

If it survives the renormalization program, then QCD is a complete and self-consistent theory without the need to introduce some extra degrees of freedom in order to generate a mass gap. We should prove that the product

$$\Delta_R^2 = Z(\lambda, \alpha, \xi, g^2)\Delta^2(\lambda, \alpha, \xi, g^2), \quad \lambda \to \infty, \quad \alpha \to 0, \tag{2.7.4}$$

where $Z(\lambda, \alpha, \xi, g^2)$ is the multiplicative renormalization constant of the mass gap, exists in the above-shown limits. However, the final result of these limits, i.e., Δ_R^2, should be achieved in the way not to compromise the general renormalizability of QCD. Contrary to the regularized version, the renormalized mass gap (2.7.4) should not depend on the gauge-fixing parameter, should be finite, positive definite, etc. Only after performing this program can one assign to it a physical meaning of a scale responsible for the true non-perturbative dynamics of QCD at large distances, in the same way as Λ_{QCD}^2 is responsible for its nontrivial perturbative dynamics at short distances. Apparently, the renormalized mass gap can be identified/related with/to the Jaffe and Witten mass gap [11]. The multiplicative renormalization program will be positively resolved in the subsequent chapters. For this we have to explicitly find $d(q^2)$ as a function of the regularized mass gap Δ^2 in the general way.

Concluding, it is worth emphasizing that the formal $\Delta^2(d) = 0$ limit implies to put the mass gap formally zero as well, namely $\Delta^2 = 0$. The arbitrary coefficient $c(\lambda, \alpha, \xi, g^2; d)$, which appears in equation (2.7.2), is, in general, not zero. In the rest of this chapter instead of the mass scale parameter $\Delta^2(d)$ we will use the mass gap Δ^2 itself, for simplicity.

2.8 Subtraction at the fundamental gluon propagator level

The non-perturbative QCD system of equations (2.6.1)–(2.6.3) depends explicitly on the mass gap Δ^2. As we already know from above, then the ghosts are not able to cancel the longitudinal component in the full gluon propagator (2.6.2). They are of no use in this case, since the transversal condition for the full gluon self-energy is always violated, see relation (2.3.20). This is the price we have paid to keep the mass gap alive in the full gluon propagator. Our aim here is to formulate a method which allows one to make the gluon propagator, relevant for the non-perturbative QCD, purely transversal in a gauge invariant way, even if the mass gap is explicitly present.

For this purpose let us define the intrinsically non-perturbative (INP) part of the full gluon propagator as follows:

$$D^{INP}_{\mu\nu}(q;\Delta^2) = D_{\mu\nu}(q;\Delta^2) - D_{\mu\nu}(q;\Delta^2 = 0) = D_{\mu\nu}(q;\Delta^2) - D^{PT}_{\mu\nu}(q), \tag{2.8.1}$$

i.e., the subtraction is made with respect to the mass gap Δ^2, and therefore the separation between these two terms is exact. Evidently, the formal $\Delta^2(D) = 0$ limit is equivalently replaced by the formal mass gap limit to zero, i.e., $\Delta^2 = 0$, as underlined above. So on account of the expressions (2.6.2) and (2.6.5), it becomes

$$\begin{aligned} D^{INP}_{\mu\nu}(q;\Delta^2) &= iT_{\mu\nu}(q)\Big[d(q^2;\Delta^2) - d^{PT}(q^2)\Big]\frac{1}{q^2} \\ &= iT_{\mu\nu}(q)d^{INP}(q^2;\Delta^2)\frac{1}{q^2}, \end{aligned} \tag{2.8.2}$$

where the explicit expression for the intrinsically non-perturbative Lorentz structure $d^{INP}(q^2;\Delta^2) = d(q^2;\Delta^2) - d^{PT}(q^2)$ can be obtained from the relations (2.6.3) and (2.6.6) for $d(q^2;\Delta^2)$ and $d^{PT}(q^2)$, respectively.

The subtraction (2.8.1) is equivalent to

$$D_{\mu\nu}(q;\Delta^2) = D^{INP}_{\mu\nu}(q;\Delta^2) + D^{PT}_{\mu\nu}(q). \tag{2.8.3}$$

The intrinsically non-perturbative gluon propagator (2.8.1) does not survive in the formal $\Delta^2 = 0$ limit. This means that it is free of the perturbative contributions, by construction. The full gluon propagator (2.6.2) in this limit is reduced to its perturbation theory counterpart (2.6.5). This means that the full gluon propagator, being also non-perturbative, nevertheless, is contaminated by the perturbative contributions. The intrinsically non-perturbative gluon propagator is purely transversal in a gauge invariant way (no special gauge choice by hand), while its full counterpart has a longitudinal component as well. There is no doubt that the true non-perturbative dynamics of the full gluon propagator is completely contained in its intrinsically non-perturbative part, since the subtraction (2.8.3) is nothing but adding zero to the full gluon propagator. We can write

$$\begin{aligned} D_{\mu\nu}(q;\Delta^2) &= i\left[T_{\mu\nu}(q)d(q^2;\Delta^2) + \xi L_{\mu\nu}(q)\right]\frac{1}{q^2} \\ &\quad - iT_{\mu\nu}(q)d^{PT}(q^2)\frac{1}{q^2} + iT_{\mu\nu}(q)d^{PT}(q^2)\frac{1}{q^2} \\ &= D^{INP}_{\mu\nu}(q;\Delta^2) + D^{PT}_{\mu\nu}(q), \end{aligned} \tag{2.8.4}$$

and so the true non-perturbative dynamics in the full gluon propagator is not affected. On the contrary it is exactly separated from its perturbative dynamics, indeed. In other words, the intrinsically non-perturbative gluon propagator is the full gluon propagator but free of its perturbative tail.

The subtraction (2.8.1) plays effectively the role of ghosts in our proposal. However, the ghosts cancel only the longitudinal component in the perturbation theory gluon propagator (2.6.5), while our proposal leads to the cancelation of the perturbative contribution in the full gluon propagator (2.6.2) as well (and thus to an automatic cancelation of its longitudinal component). Nevertheless, this is not a problem, since the mass gap does not survive in the formal perturbation theory limit anyway. The only problem with it is that, being exact, formally it may not be unique [41, 42]. However, the uniqueness of such kinds of separation can be achieved only in the explicit solution for the full gluon propagator as a function of the mass gap (see next Chapters). Anyway, this subtraction is a first necessary step, which guarantees the transversality of the intrinsically non-perturbative gluon propagator $D^{INP}_{\mu\nu}(q;\Delta^2)$ without losing even one bit of the information on the true non-perturbative dynamics in the full gluon propagator $D_{\mu\nu}(q;\Delta^2)$. At the same time, its non-trivial perturbative dynamics is completely saved in its perturbation theory gluon propagator $D^{PT}_{\mu\nu}(q)$. It is worth emphasizing that both terms in the subtraction (2.8.3) are valid in the whole momentum range; they are not asymptotics.

The full gluon propagator (2.6.2), by keeping the mass gap alive, is not physical in the sense that it cannot be made transversal by ghosts. Therefore it cannot be used for numerical calculations of the physical observables from first principles. However, our suggestion makes it possible to present $D_{\mu\nu}(q;\Delta^2)$ as the exact sum (2.8.3) of the two physical propagators. The intrinsically non-perturbative gluon propagator $D^{INP}_{\mu\nu}(q;\Delta^2)$ is automatically transversal, by construction. It fully contains all the information of the full gluon propagator on its non-perturbative context. On the other hand, the perturbation theory gluon propagator $D^{PT}_{\mu\nu}(q)$ is free of the mass gap, and hence the ghosts will cancel its longitudinal component, making it thus transversal (physical). It fully contains all the information of the full gluon propagator on its non-trivial perturbative context (scale violation, asymptotic freedom).

Let us show explicitly the Schwinger–Dyson equation for the intrinsically non-perturbative gluon propagator. From equation (2.6.1) and (2.6.4), on account of the subtraction (2.8.3), one gets

$$D^{INP}_{\mu\nu}(q;\Delta^2) = D^0_{\mu\rho}(q) iT_{\rho\sigma}(q)\Delta^2 D^{PT}_{\sigma\nu}(q) \\ + D^0_{\mu\rho}(q) iT_{\rho\sigma}(q)\left[q^2\Pi(q^2;D) + \Delta^2\right] D^{INP}_{\sigma\nu}(q;\Delta^2), \quad (2.8.5)$$

where $D^{PT}_{\sigma\nu}(q)$ satisfies its own equation (2.6.4).

The remarkable feature of this equation is that, by switching interaction formally off (i.e., setting $\Pi(q^2;D) = \Delta^2 = 0$), it cannot be reduced to the free gluon propagator, like occurs for the full gluon propagator (2.6.1) and its perturbation theory counterpart (2.6.4). In other words, in the intrinsically non-perturbative QCD the gluon propagator is always dressed. And this brings in one more serious argument in favor of the above proposed subtractions of all the types of the perturbative contributions. The emission and absorbtion of the colored dressed gluons at large distances can be suppressed by the renormalization of the mass gap (see Chapters below). At the same time, there exists no such mechanism to do the same with the colored free gluons in order to ensure their confinement. So the correct theory of low-energy QCD should exclude the free gluon propagator from its formalism. This just takes place in the intrinsically non-perturbative QCD. Both the suppression of the dressed gluons and the absence of the free gluons are necessary for the explanation of gluon confinement by the intrinsically non-perturbative QCD. On the other hand, the full gluon propagator (2.6.2) which satisfies equation (2.6.1) is reduced to the free gluon propagator when the interaction is formally switched off. There is no mechanism to suppress the emission and absorbtion of the free gluons at large distances. That is why the full gluon propagator (2.6.2) is not confining, while the intrinsically non-perturbative one can confine (see next Chapters).

The subtraction (2.8.3) *seems to be necessary, indeed. It makes the relevant gluon propagator transversal and excludes the free gluons from the theory at the same time.*

Concluding, the solution of the above-mentioned two problems on how to preserve the color gauge invariance/symmetry in QCD at non-zero mass gap completes our investigation in this chapter. This means that from now on we can forget the relation (2.3.20) completely, since there are no longer any negative consequences for the intrinsically non-perturbative QCD. In this connection let us underline that all the initial subtractions have been done in a gauge invariant way (i.e., not in the separate propagators, which enter the skeleton loop integrals, contributing to the full gluon self-energy). The same is true for the subtraction with respect to the mass gap in equation (2.8.3).

2.9 Discussion

The mass scale parameter $\Delta^2(D) = \Delta_t^2(D)$ is nothing but the constant skeleton tadpole term, namely

$$\Pi_t(D) = \Delta_t^2(D) = g^2 \int \frac{id^4 q_1}{(2\pi)^4} T_4^0(q_1, 0, 0, -q_1) D(q_1), \qquad (2.9.1)$$

where we omit the tensor and color indices, for simplicity. It is generated by the interactions of massless gluon modes in the gluon sector of QCD. All the full gluon vertices are non-explicitly involved through the dependence on D. But the point-like four-gluon vertex explicitly plays a key role in the dynamical generation of the mass gap, introduced in the relation (2.7.2). The dominant role of the point-like four-gluon vertex is due to the two facts: the triple gluon vertex vanishes when all the gluon momenta involved go to zero ($T_3(0,0) = 0$), while its four-gluon counterpart survives ($T_4(0,0,0) \neq 0$). In this connection let us remind that the mass gap determines the structure of the full gluon Schwinger – Dyson equation (2.6.2) and its solution (2.6.3) just in this limit. So there is no doubt in the important role of the quartic gluon vertex in the generation of the mass gap [33] (see also general discussion in [11]). It has not been introduced by hand, since through the tadpole term it is explicitly present in the sum of the skeleton loop integrals, contributing to the full gluon self-energy (2.2.3).

All the quantities considered in this Chapter are necessarily regularized, as a first step. However, nothing depends on our approach on the specific regularization scheme, preserving or not gauge invariance. It is impossible to perform any concrete calculations of the regularized skeleton loop integrals, containing unknown, in general, the full propagators and vertices. No truncations/approximations/assumptions (which means no use of perturbation theory), special gauge choice, etc., have been made for them. Only algebraic derivations have been done so far.

The mass gap Δ^2 violates explicitly the Slavnov – Taylor identity for the full gluon propagator, which satisfies the corresponding equation of motion. Also, in its presence the ghosts are not able to cancel the longitudinal component in the full gluon propagator in order to guarantee unitarity of the S-matrix in this theory. So it should be disregarded on general grounds, put formally zero everywhere. We have explicitly shown that this formal limit is equivalent to the perturbative $q^2 \to \infty$ limit and *vice versa*, which leads to the formulation of the system of equations for the perturbative QCD.

In order to confirm that the color gauge invariance/symmetry of QCD is maintained in the explicit presence of the mass gap as well, we have introduced the spurious mechanism. It makes it possible to modify the original gluon Schwinger – Dyson equation for the full gluon propagator in a such way that makes the Slavnov – Taylor identity automatically satisfied at non-zero Δ^2. At the same time, the dynamical context of the modified gluon Schwinger – Dyson equation is not affected, it is the same as the original gluon Schwinger – Dyson equation. The solution (2.6.3) depends regularly on the mass gap, and it has a correct perturbation theory $\Delta^2 = 0$ limit, shown in the solution (2.6.6). From them it clearly follows that the effect of the mass gap dominates at $q^2 \to 0$ and is strongly suppressed in the perturbative $q^2 \to \infty$ regime, so it is justified to simply drop it in the perturbation theory.

Our solution (2.6.3) depends explicitly on the mass gap. As underlined above, in this case the ghosts are of no use to cancel the longitudinal component in the full gluon propagator. However, we have formulated a general method which makes it possible to achieve the transversality of the full gluon propagator, relevant for the non-perturbative QCD, in a gauge invariant way. It is based on the exact subtraction of the perturbative contribution from the full gluon propagator. The obtained intrinsically non-perturbative gluon propagator is purely transversal, maintaining thus unitarity of the S-matrix within our approach. It completely reproduces the true non-perturbative structure of the full gluon propagator, and in parallel, it is free of perturbative contaminations at the fundamental gluon level.

As pointed out above, we need no ghosts to ensure the cancelation of the longitudinal component in the full gluon propagator. Nevertheless, this does not mean that we need no ghosts at all. We need them in other sectors of QCD, for example in the quark Slavnov – Taylor identity, which contains the so-called ghost–quark scattering kernel explicitly [2]. This kernel still makes an important contribution to the identity even if the relevant gluon propagator is transversal [20, 22, 34, 35]. Do not mix the intrinsically non-perturbative gluon propagator (2.8.2) with the full gluon propagator (2.6.2) in the Landau gauge. The former is transversal by construction in a gauge invariant way. The latter one becomes transversal only by choosing the Landau gauge by hand, i.e., not in a gauge invariant way (let us remind that the ghosts cannot make it transversal).

In the place of the mass gap, any other mass scale parameter might serve. This could be introduced into the full gluon propagator by hand, as an ansatz, or arise as a result of some specific approximation/truncation/ assumption made in the gluon Schwinger – Dyson equation itself and hence in its solution, etc. Its origin is irrelevant for our method, the only request is that the full gluon propagator should regularly depend on it. However, none of the truncations/approximations/assumptions or ansatzs made or introduced in the framework of any approach should undermine the above-discussed general role of ghosts in perturbative QCD. Our method just guarantees this.

Evidently, nothing will be changed in our final expressions here and below if one formally includes the mass scale parameters Δ_q^2 and $\Delta_g^2(D)$ into the tadpole term, thus, replacing $\Delta_t^2(D) \equiv \Delta^2(D) \to \Delta^2(D) = \Delta_t^2(D) + \Delta_q^2 + \Delta_g^2(D)$. In this case $\Delta^2(D)$ has to be treated as the re-defined tadpole term, while keeping both transversality conditions (2.3.3) and (2.3.11) fulfilled.

The common belief (which comes from the perturbation theory) that the tadpole term contradicts the color gauge invariance/symmetry of QCD is false. This fundamental symmetry is maintained/preserved at non-zero tadpole term as well. This means that the mass gap itself exists in QCD. In fact, the mass gap is generated by the tadpole term: without the tadpole term there is no the mass gap [36–38].

2.A Appendix: Application for Abelian case

It is instructive to discuss in more detail why the mass gap does not occur in Quantum Electrodynamics. The equation for the full photon propagator $D(q)$ can be symbolically written down as follows:

$$D(q) = D_0(q) + D_0(q)\Pi(q)D(q), \tag{2.A.1}$$

where we omit, for convenience, the dependence on the Dirac indices, and $D_0(q)$ is the free photon propagator. $\Pi(q)$ describes the vacuum polarization tensor contribution to the photon self-energy. Analytically it looks

$$\Pi(q) \equiv \Pi_{\mu\nu}(q) = -g^2 \int \frac{i\, d^4p}{(2\pi)^4} Tr\left[\gamma_\mu S(p-q)\Gamma_\nu(p-q,q)S(p)\right], \tag{2.A.2}$$

where $S(p)$ and $\Gamma_\mu(p-q,q)$ represent the full electron propagator and the full electron–photon vertex, respectively. Here and everywhere below the signature again is Euclidean, since it implies $q_i \to 0$ when $q^2 \to 0$ and *vice versa*. This tensor has the dimensions of mass squared, and therefore it is quadratically divergent. It should be regularized.

Similar to the quark case in Section 2.3, let us introduce the mass gap through the definition of the subtracted photon self-energy as follows:

$$\Pi^s(q) \equiv \Pi^s_{\mu\nu}(q) = \Pi_{\mu\nu}(q) - \Pi_{\mu\nu}(0) = \Pi_{\mu\nu}(q) - \delta_{\mu\nu}\Delta^2(\lambda), \tag{2.A.3}$$

where the dimensionless ultraviolet regulating parameter λ has been introduced into the mass gap $\Delta^2(\lambda)$, given by the integral (2.A.2) at $q^2 = 0$, in order to assign a mathematical meaning to it. In this connection let us note that what we have said about the regularization of all the quantities in Section 2.2 is valid here as well, apart from one observation. The above-mentioned subtraction at zero point $q^2 = 0$ is not dangerous in QED, since it is an abelian quantum gauge theory. In this theory there is no self-interaction of massless photons, which may be the source of the singularities in the $q^2 \to 0$ limit.

The decompositions of the vacuum polarization tensor and its subtracted counterpart into the independent tensor structures can be written as follows:

$$\begin{aligned} \Pi_{\mu\nu}(q) &= T_{\mu\nu}(q)q^2\Pi_1(q^2) + q_\mu q_\nu(q)\Pi_2(q^2), \\ \Pi^s_{\mu\nu}(q) &= T_{\mu\nu}(q)q^2\Pi^s_1(q^2) + q_\mu q_\nu(q)\Pi^s_2(q^2), \end{aligned} \tag{2.A.4}$$

where all the invariant functions of q^2 are dimensionless ones. In addition, $\Pi^s_n(q^2)$ at $n = 1, 2$ cannot have the pole-type singularities in the $q^2 \to 0$ limit, since $\Pi^s(0) = 0$, by definition; otherwise all the invariant functions remain arbitrary. From these relations it follows that $\Pi^s(q) = O(q^2)$, thus, it is always of the order q^2.

Substituting these decompositions into subtraction (2.A.3), one obtains

$$\Pi_2\left(q^2\right) = \Pi_2^s\left(q^2\right) + \frac{\Delta^2(\lambda)}{q^2},$$
$$\Pi_1\left(q^2\right) = \Pi_1^s\left(q^2\right) + \frac{\Delta^2(\lambda)}{q^2}. \tag{2.A.5}$$

On the other hand from the transversality condition for the photon self-energy

$$q_\mu \Pi_{\mu\nu}(q) = q_\nu \Pi_{\mu\nu}(q) = 0, \tag{2.A.6}$$

which comes from the current conservation in QED, one arrives at the $\Pi_2(q^2) = 0$. Then from the relation (2.A.5) it follows that

$$\Pi_2^s\left(q^2\right) = -\frac{\Delta^2(\lambda)}{q^2}, \tag{2.A.7}$$

which, however, is impossible since $\Pi_2^s(q^2)$ cannot have a massless single particle singularity, as mentioned above. So the mass gap should be discarded, or put formally zero and, consequently, $\Pi_2^s(q^2)$ as well. Setting

$$\Delta^2(\lambda) = 0, \quad \text{and} \quad \Pi_2^s\left(q^2\right) = 0. \tag{2.A.8}$$

Thus the subtracted photon self-energy is also transversal and satisfies the transversality condition $q_\mu \Pi^s_{\mu\nu}(q) = q_\nu \Pi^s_{\mu\nu}(q) = 0$. This means that it coincides with the photon self-energy. This comes from the subtraction (2.A.3), on account of the relations (2.A.8),

$$\Pi_{\mu\nu}(q) = \Pi^s_{\mu\nu}(q) = T_{\mu\nu}(q) q^2 \Pi_1^s\left(q^2\right), \tag{2.A.9}$$

so that the photon self-energy does not have a pole in its invariant function $\Pi_1(q^2) = \Pi_1^s(q^2)$. In obtaining these results neither was the perturbation theory used nor a special gauge chosen. So there is no place for a quadratically divergent constant in QED, while logarithmic divergence still can be present in the invariant function $\Pi_1\left(q^2\right) = \Pi_1^s\left(q^2\right)$. It is to be included into the electric charge through the corresponding renormalization program — for these detailed gauge-invariant derivations explicitly done in lower order of the perturbation theory see, for example, Refs. [28, 183, 237].

In fact, the current conservation condition (2.A.6), thus, transversality of the photon self-energy lowers the quadratic divergence of the corresponding integral (2.A.2) to a logarithmic one. That is the reason why in QED only logarithmic divergences survive. The current conservation condition for the photon self-energy (2.A.6) and the Slavnov–Taylor identity for the full photon propagator $q_\mu q_\nu D_{\mu\nu}(q) = i\xi$ are consequences of gauge invariance. They should be maintained at every stage of the calculations, since the photon is a physical state. In other words, at all stages the current conservation plays a crucial role in extracting physical information from the S-matrix elements in QED, which are usually proportional to the combination $j_1^\mu(q) D_{\mu\nu}(q) j_2^\nu(q)$. The current conservation condition $j_1^\mu(q) q_\mu = j_2^\nu(q) q_\nu = 0$ implies that the unphysical (longitudinal) component of the full photon propagator does not change the physics of QED, so only its physical (transversal) component is important. In its turn this means that the transversality condition imposed on the photon self-energy is also important, because $\Pi_{\mu\nu}(q)$ itself is a correction to the amplitude of the physical process, for example such as electron–electron scattering.

Thus in QED there is no mass gap, and we can replace $\Pi(q)$ by its subtracted counterpart $\Pi^s(q)$ from the very beginning, $\Pi(q) \to \Pi^s(q)$, totally discarding the quadratically divergent constant $\Delta^2(\lambda)$ from all the equations and relations. Then the photon equation (2.A.1) becomes

$$D(q) = D_0(q) + D_0(q)\Pi^s(q)D(q), \tag{2.A.10}$$

which is equivalent to

$$D(q) = \frac{D_0(q)}{1 - \Pi^s(q)D_0(q)}, \tag{2.A.11}$$

and this is the summation of geometric series, namely

$$D(q) = D_0(q) + D_0(q)\Pi^s(q)D_0(q) - D_0(q)\Pi^s(q)D_0(q)\Pi^s(q)D_0(q) + \dots. \tag{2.A.12}$$

Since $\Pi^s(q) = O(q^2)$ and $D_0(q) \sim (q^2)^{-1}$, the infrared singularity of the full photon propagator is determined by the infrared singularity of the free photon propagator, which is always $1/q^2$. Hence the photons even if dressed always remain massless.

Concluding, we cannot release the mass gap from the QED vacuum, while we can release the photons and the electron–positron pairs from it. In QCD the situation is completely opposite to QED. In this theory we can release the mass gap from its vacuum, as it has been described in this book. But we cannot release the gluons and the quark/antiquark pairs from the QCD vacuum because of the color confinement phenomenon. When we speak about the mass gap to be released or not from the vacua of the corresponding quantum gauge theories, we mean, of course, whether it should be discarded or not from the very beginning in these theories. In this connection, let us remind that at this stage the mass gap is not physical, it is only a regularized quantity.

Problems

Problem 2.1. *Prove on general grounds that* $D_{\mu\nu}(q)D^{-1}_{\mu\alpha}(q) = \delta_{\nu\alpha}$.

Problem 2.2. *What are the inverse of the full* (2.4.2) *and free* (2.2.2) *gluon propagators? For the solution use the results of the Problem 2.1.*

Problem 2.3. *Prove on a general level that both invariant functions for the subtracted quark part in the relations* (2.3.5) *are the regular functions of* q^2 *in the* $q^2 \to 0$ *limit.*

Problem 2.4. *Derive the relations given by equation* (2.3.6).

Problem 2.5. *Derive equation* (2.3.26).

Problem 2.6. *Derive the relation given by equation* (2.4.4).

Problem 2.7. *Derive the equation* (2.5.3) *and the expression* (2.5.4).

Problem 2.8. *Derive explicit expressions* (2.6.3) *and* (2.6.6).

Problem 2.9. *Derive the equation* (2.8.2) *by substituting equation* (2.6.1) *and* (2.6.4). *Use for this expression* (2.2.2) *and equations* (2.6.2) *and* (2.6.5).

Problem 2.10. *Derive the equation* (2.8.5).

Problem 2.11. *What is the solution of equation* (2.8.5) *in terms of* $d^{INP}(q^2)$ *given by the equation* (2.8.2)*?*

Chapter 3

Formal Exact Solutions for the Full Gluon Propagator at Non-zero Mass Gap

3.1 Introduction

Our primary goal in this Chapter is to find formal exact solutions for the full gluon propagator as a function of the regularized mass gap. However, for the sake of completeness and further clarity, it is instructive to describe briefly here the main results of the previous Chapter.

The first problem (i), formulated in Section 2.5, namely how to preserve the color gauge invariance in QCD at non-zero mass gap, has been solved by introducing the spurious technique. It made it possible to show that the Slavnov – Taylor identity (2.4.1) can be automatically satisfied at non-zero mass gap Δ^2 as well. The corresponding Schwinger – Dyson equation for the full gluon propagator has been also derived, the system of equations (2.6.1)–(2.6.3). For convenience, let us show it explicitly once more

$$D_{\mu\nu}(q) = D^0_{\mu\nu}(q) + D^0_{\mu\rho}(q) i T_{\rho\sigma}(q) \left[q^2\Pi(q^2; D) + \Delta^2(D)\right] D_{\sigma\nu}(q), \quad (3.1.1)$$

where

$$D_{\mu\nu}(q) = i \left[T_{\mu\nu}(q) d(q^2) + \xi L_{\mu\nu}(q)\right] \frac{1}{q^2}, \quad (3.1.2)$$

and its invariant function is given by

$$d(q^2) = \frac{1}{1 + \Pi(q^2; D) + (\Delta^2(D)/q^2)}. \quad (3.1.3)$$

In these equations the dependence of the full gluon propagator on the mass scale parameter $\Delta^2(D)$ was not explicitly shown, for simplicity. Here it is important to remind that the gluon invariant function $\Pi(q^2; D)$ does not depend on $\Delta^2(D)/q^2$ at all, it is a regular function at $q^2 \to 0$. At $q^2 \to \infty$ limit it is logarithmically divergent: otherwise it remains arbitrary (it is a sum of the corresponding regularized skeleton loop integrals, contributing to the subtracted full gluon self-energy). The free gluon propagator $D^0_{\mu\nu}(q)$ is to be obtained from the decomposition (3.1.2) by putting $d(q^2) = 1$. It is also worth emphasizing that the spurious mechanism did not affect the dynamical context of the original gluon Schwinger–Dyson equation. In other words, it makes it possible to retain the mass gap in the transversal part of the gluon Schwinger–Dyson equation and, at the same time, to cancel the term in its longitudinal part, which violates the Slavnov–Taylor identity.

By the mass gap we understand some fixed mass squared which is related to $\Delta^2(D)$ as follows, see equation (2.7.2),

$$\Delta^2(D) = \Delta^2 \cdot c(D), \tag{3.1.4}$$

where the dimensionless constant $c(D)$ depends on D, while the fixed mass squared Δ^2 does not depend on D. We call it the mass gap.

The second problem (ii), formulated in Section 2.5, namely how to make the gluon propagator, relevant for the non-perturbative QCD, purely transversal in a gauge invariant way, has been solved by the proposed subtraction in equation (2.8.1). The non-perturbative system of equations (3.1.1)–(3.1.4) depends explicitly on the mass gap Δ^2. Then the ghosts are not able to cancel the longitudinal component in the full gluon propagator — they are of no use in this case. The above-mentioned subtraction is equivalent to

$$D_{\mu\nu}(q;\Delta^2) = D^{INP}_{\mu\nu}(q;\Delta^2) + D^{PT}_{\mu\nu}(q). \tag{3.1.5}$$

The full gluon propagator (3.1.1) and hence (3.1.5), keeping the mass gap alive, is not physical in the sense that it cannot be made transversal by ghosts. Therefore it cannot be used for numerical calculations of the physical observables from first principles. However, our suggestion makes it possible to present it as the exact sum of the two physical propagators. The intrinsically non-perturbatiive gluon propagator $D^{INP}_{\mu\nu}(q;\Delta^2)$ is automatically transversal, see equation (2.8.2). It fully contains all the information of the full gluon propagator on its non-perturbative dynamics. Just it should be used in order to calculate the physical observables in low-energy QCD. In high-energy QCD the perturbation theory gluon propagator $D^{PT}_{\mu\nu}(q)$ is to be used. It is free of the mass gap and the ghosts can cancel its longitudinal component, making thus it transversal/physical.

Concluding, we have briefly recalled how to preserve the color gauge invariance/symmetry in QCD at non-zero mass gap. Our approach makes it possible to save the mass gap in the transversal part of the gluon

Schwinger – Dyson equation, and at the same to satisfy the Slavnov – Taylor identity. It also makes the gluon propagator, relevant for the non-perturbative QCD, purely transversal in a gauge invariant way when the mass gap is explicitly present.

3.2 Singular solution

In order to find the general formal solution for the full gluon propagator, let us begin with equation (3.1.3) which is convenient to re-write as follows:

$$d(q^2) \equiv d(q^2;\Delta^2) = \frac{1}{1+\Pi(q^2;d)+c(d)(\Delta^2/q^2)}, \tag{3.2.1}$$

where the dependence on D is replaced by the equivalent dependence on d, and the relation (3.1.4) is already used. Let us introduce further the dimensionless variable $z = \Delta^2/q^2$. The full Lorentz structure (3.2.1) regularly depends on the mass gap, and hence on z. Then it can be formally expanded in a Taylor series in powers of z around zero z as follows:

$$d(q^2;\Delta^2) = d(q^2;z) = \sum_{k=0}^{\infty} z^k f_k(q^2), \tag{3.2.2}$$

where the functions $f_k(q^2)$ are the corresponding derivatives of $d(q^2;z)$ with respect to z at $z=0$, which is equivalent to the perturbation theory $\Delta^2=0$ limit. In this connection, let us repeat once more that in this limit d becomes d^{PT}, and the invariant function $\Pi(q^2;d)$ does not depend on the variable z, i.e., in fact, $\Pi(q^2;d) \equiv \Pi(q^2;d^{PT})$. From the expression (3.2.1) one obtains

$$f_k(q^2) = (-1)^k d^{PT}(q^2)\left[d^{PT}(q^2)c(d^{PT})\right]^k, \tag{3.2.3}$$

where $d(q^2;z=0) = d^{PT}(q^2)$ and $c(d) = c(d^{PT})$ at $\Delta^2 = 0$, by definition. It is worth emphasizing that, contrary to the relation (3.2.1), the expansion (3.2.2) can be considered now as a formal solution for $d(q^2)$, since $f_k(q^2)$ depend on $d^{PT}(q^2)$, which is assumed to be known.

The functions $f_k(q^2)$ are regular functions of the variable q^2, since they finally depend on $d^{PT}(q^2)$ which is a regular function of q^2. Therefore they can be expanded in a Taylor series near $q^2=0$ (here we can put the subtraction point $\alpha=0$, for simplicity, since all the quantities are already regularized, they depend on α and so on). Introducing the dimensionless variable $x = q^2/M^2$, where M^2 is some auxiliary mass squared, it is convenient to present this expansion as a sum of the two terms, namely

$$f_k(q^2) = \sum_{n=0}^{k} x^n f_{kn}(0) + x^{k+1}B_k(x), \tag{3.2.4}$$

where the coefficient $f_{kn}(0)$ are the corresponding derivatives of the functions $f_k(q^2) \equiv f_k(x)$ with respect to x at $x = 0$. Of course, these coefficients depend on the parameters of the theory such as $\lambda, \alpha, \xi, g^2$, and so on, which are not shown explicitly. The dependence on these parameters will be restored at the final stage of our derivations. The dimensionless functions $B_k(x)$ are regular functions of x; otherwise they remain arbitrary.

So the general Lorentz structure (3.2.2) becomes

$$d(q^2) = \sum_{k=0}^{\infty} z^k f_k(x) = \sum_{k=0}^{\infty} z^k \left(\sum_{n=0}^{k} x^n f_{kn}(0) + x^{k+1} B_k(x) \right). \qquad (3.2.5)$$

Omitting all the intermediate tedious derivations, these double sums can be equivalently presented as the sum of the three independent terms as follows:

$$d(q^2) = z \sum_{k=0}^{\infty} z^k \sum_{m=0}^{\infty} \Phi_{km}(0) + a \sum_{k=0}^{\infty} a^k \sum_{m=0}^{\infty} A_{km}(x) + d^{PT}(q^2), \qquad (3.2.6)$$

where the constant $a = xz = \Delta^2/M^2$ and the dimensionless functions $A_{km}(x)$ are regular functions of x: otherwise they remain arbitrary. $d^{PT}(q^2)$ denotes the terms which do not depend on the mass gap Δ^2 at all — which is nothing but the Lorentz structure of the perturbation theory gluon propagator (2.6.6), indeed.

Going back to the gluon momentum variable q^2, one obtains

$$d(q^2; \Delta^2) = d^{INP}(q^2; \Delta^2) + d^{MPT}(q^2; \Delta^2) + d^{PT}(q^2), \qquad (3.2.7)$$

where the superscript "MPT" stands for the mixed perturbation theory part of the full effective charge $d(q^2; \Delta^2)$. Their explicit expressions are

$$\begin{aligned} d^{INP}(q^2; \Delta^2) &= \left(\frac{\Delta^2}{q^2}\right) \sum_{k=0}^{\infty} \left(\frac{\Delta^2}{q^2}\right)^k \Phi_k \\ &= \left(\frac{\Delta^2}{q^2}\right) \sum_{k=0}^{\infty} \left(\frac{\Delta^2}{q^2}\right)^k \sum_{m=0}^{\infty} \Phi_{km} \end{aligned} \qquad (3.2.8)$$

and

$$\begin{aligned} d^{MPT}(q^2; \Delta^2) &= \left(\frac{\Delta^2}{M^2}\right) \sum_{k=0}^{\infty} \left(\frac{\Delta^2}{M^2}\right)^k A_k(q^2) \\ &= \left(\frac{\Delta^2}{M^2}\right) \sum_{k=0}^{\infty} \left(\frac{\Delta^2}{M^2}\right)^k \sum_{m=0}^{\infty} A_{km}(q^2), \end{aligned} \qquad (3.2.9)$$

and both expansions are expansions in powers of the mass gap. Here and everywhere below all the quantities depend on the parameters of the theory, namely $\Delta^2 = \Delta^2(\lambda, \alpha, \xi, g^2)$ and $A_k(q^2) = \sum_{m=0}^{\infty} A_{km}(q^2; \lambda, \alpha, \xi, g^2)$. At the same time, Φ_{km} depends in addition on the parameter a as well, i.e., $\Phi_{km} = \Phi_{km}(\lambda, \alpha, \xi, g^2, a)$.

The exact structure of the singular solution

In order to eliminate the dependence of such obtained expressions on the arbitrariness of the auxiliary mass scale parameter M^2, one should go to the $a = \Delta^2/M^2 \to 0$ limit, which is equivalent to the $M^2 \to \infty$ limit at a fixed mass gap. Since the functions $A_{km}(q^2)$ in the mixed perturbation theory part (3.2.9) are regular functions of q^2, this term simply vanishes in this limit. The full gluon propagator (3.1.2) thus finally becomes the sum of the two independent terms in complete agreement with (3.1.5), namely

$$D_{\mu\nu}(q;\Delta^2) = D_{\mu\nu}^{INP}(q;\Delta^2) + D_{\mu\nu}^{PT}(q), \tag{3.2.10}$$

where

$$D_{\mu\nu}^{INP}(q;\Delta^2) = iT_{\mu\nu}(q)\frac{\Delta^2}{(q^2)^2}L(q^2;\Delta^2) \tag{3.2.11}$$

with

$$L(q^2;\Delta^2) = \sum_{k=0}^{\infty}\left(\frac{\Delta^2}{q^2}\right)^k \Phi_k = \sum_{k=0}^{\infty}\left(\frac{\Delta^2}{q^2}\right)^k \sum_{m=0}^{\infty} \Phi_{km}, \tag{3.2.12}$$

and now Φ_{km} do not depend on a, i.e., $\Phi_{km} = \Phi_{km}(\lambda, \alpha, \xi, g^2)$, while

$$D_{\mu\nu}^{PT}(q) = i\left[T_{\mu\nu}(q)d^{PT}(q^2) + \xi L_{\mu\nu}(q)\right]\frac{1}{q^2} \tag{3.2.13}$$

with $d^{PT}(q^2)$ given in equation (2.6.6).

The general problem of convergence of the formal (but regularized) series, which appear in these relations, is irrelevant here. In other words, it does not make any sense to discuss the convergence of such kinds of series before the renormalization program is performed — which will allow one to see whether or not the mass gap survives it at all. The problem of how to remove the ultraviolet overlapping divergences [15] and usual overall ones [2–5] is a standard one. However, it is not our problem since, let us remind that, the mass gap does not survive in the perturbation theory at the $q^2 \to \infty$ limit. Our problem here will be how to deal with severe infrared ($q^2 \to 0$) singularities due to their novelty and genuine (intrinsic) non-perturbative character. In this limit the mass gap dominates the structure of the full gluon propagator. Fortunately, there already exists a well-elaborated mathematical formalism for this purpose, namely the distribution theory [40], into which the dimensional regularization method [29] should be correctly implemented (see also Appendix 3.A and 3.B).

Let us now remind that the full gluon propagator (3.2.10), depending on the mass gap, is not physical. It cannot be made transversal by ghosts. At the same time, the intrinsically non-perturbative gluon propagator $D^{INP}_{\mu\nu}(q;\Delta^2)$ is automatically transversal, by its construction, and fully represents the true non-perturbative dynamics of the full gluon propagator. Taking this observation into account, we propose to replace

$$D_{\mu\nu}(q;\Delta^2) \rightarrow D^{INP}_{\mu\nu}(q;\Delta^2) = D_{\mu\nu}(q;\Delta^2) - D^{PT}_{\mu\nu}(q), \tag{3.2.14}$$

and consider it as the relevant gluon propagator in order to calculate the physical observables in low-energy QCD from first principles.

On the other hand, the perturbation theory gluon propagator $D^{PT}_{\mu\nu}(q)$ is free of the mass gap, and hence the ghosts will cancel its longitudinal component, making it transversal/physical. It fully contains all the information of the full gluon propagator on its non-trivial perturbative context (scale violation, asymptotic freedom, etc.). Taking this observation into account, we propose to replace

$$D_{\mu\nu}(q;\Delta^2) \rightarrow D^{PT}_{\mu\nu}(q) = D_{\mu\nu}(q;\Delta^2) - D^{INP}_{\mu\nu}(q;\Delta^2), \tag{3.2.15}$$

and consider it as the relevant gluon propagator in order to calculate the physical observables in high-energy QCD from first principles.

It is worth reminding once more that both terms which appear in the right-hand-side of equation (3.2.10) are valid in the whole energy/momentum range, thus they are not asymptotics. At the same time, we have achieved the exact and unique (see discussion below) separation between the terms responsible for the intrinsically non-perturbative (dominating in the infrared at $q^2 \rightarrow 0$) and the nontrivial perturbation theory (dominating in the ultraviolet at $q^2 \rightarrow \infty$) dynamics in the true QCD vacuum. The structure of this solution shows clearly that the deep infrared region interesting for confinement and other non-perturbative effects is dominated by the mass gap. In the formal perturbation theory, $\Delta^2 = 0$ limit, the nontrivial perturbation theory dynamics is all that matters.

Intrinsically non-perturbative part of the singular solution

In accordance with our prescription, one should subtract all the types of the perturbation theory contributions in order to get the relevant gluon propagator for the intrinsically non-perturbative QCD. As it follows from the discussion above, in the case of the general singular solution, we should subtract the perturbation theory term only. Doing so in equation (3.2.14), one finally obtains

$$D^{INP}_{\mu\nu}(q;\Delta^2) = iT_{\mu\nu}(q)d^{INP}(q^2;\Delta^2)\frac{1}{q^2}, \tag{3.2.16}$$

where

$$d^{INP}(q;\Delta^2) = \frac{\Delta^2}{q^2}L(q^2;\Delta^2) = \frac{\Delta^2}{q^2}\sum_{k=0}^{\infty}\left(\frac{\Delta^2}{q^2}\right)^k \Phi_k, \qquad (3.2.17)$$

and here $\Delta^2 = \Delta^2\left(\lambda, \alpha, \xi, g^2\right)$, while $\Phi_k = \Phi_k(\lambda, \alpha, \xi, g^2) = \sum_{m=0}^{\infty} \Phi_{km}(\lambda, \alpha, \xi, g^2)$.

The intrinsically non-perturbative part (3.2.16) of the full gluon propagator is characterized by the presence of severe power-type (or, equivalently, non-perturbative) infrared singularities $(q^2)^{-2-k}$, $k = 0, 1, 2, 3, \ldots$. So these singularities are defined as more singular than the power-type infrared singularity of the free gluon propagator $(q^2)^{-1}$, which thus can be defined as the perturbation theory infrared singularity. The intrinsically non-perturbative part of the full gluon propagator, apart from the structure (Δ^2/q^4), is nothing but the corresponding Laurent expansion, explicitly shown in equation (3.2.17), in integer powers of q^2 accompanied by the corresponding powers of the mass gap squared and multiplied by the q^2-independent factors, the so-called residues $\Phi_k(\lambda, \alpha, \xi, g^2) = \sum_{m=0}^{\infty} \Phi_{km}(\lambda, \alpha, \xi, g^2)$. The sum over m indicates that an infinite number of the contributions of the above-mentioned corresponding regularized skeleton loop integrals invokes each severe infrared singularity labeled by k. It is worth emphasizing that the Laurent expansion given by (3.2.17) cannot be summed up into some known function, since its residues are, in general, arbitrary. However, this arbitrariness is not a problem. The functional dependence, which has been established exactly, is all that matters. Let us note that the expansion has been independently obtained in [39] in a rather different way.

It is important to emphasize that the intrinsically non-perturbative gluon propagator (3.2.16) is *uniquely* defined because there exists a special regularization expansion for severe, non-perturbative, infrared singularities, while for the perturbation theory infrared singularity such a kind of expansion does not exist at all. See Appendix 3.B and [27, 39, 40]. This just determines the principal difference between them. It is also *exactly* defined because of its two features:

- The first one is that, the intrinsically non-perturbative gluon propagator depends only on the transversal degrees of freedom of gauge bosons.
- The second one is that in the formal perturbation theory $\Delta^2 = 0$ limit the intrinsically non-perturbative gluon propagator vanishes.

Thus, one can conclude that the presence of severe infrared singularities only is the first necessary condition, while the regular dependence on the mass gap and the transversality are only second sufficient conditions for the unique and exact separation of the intrinsically non-perturbative gluon propagator from the perturbation theory gluon propagator. In other words, the intrinsically non-perturbative gluon propagator is free of all the types of the perturbation theory contributions (contaminations). It should replace the full gluon propagator in order to calculate the physical observables, processes, etc. from first principles in low-energy QCD after the corresponding renormalization program is performed. It is worth reminding that the intrinsically non-perturbative gluon propagator satisfies its own equation of motion (2.8.5).

3.3 Massive solution

One of the direct consequences of the explicit presence of the mass gap in the full gluon propagator is that the gluon may acquire an effective mass, indeed [43]. From equation (3.2.1) it follows that

$$\frac{1}{q^2}d(q^2) = \frac{1}{q^2 + q^2\Pi(q^2;\xi) + \Delta^2 c(\xi)}, \tag{3.3.1}$$

where instead of the dependence on D the dependence on ξ is explicitly shown, while here and below the dependence on all other parameters is not shown, for simplicity. The full gluon propagator (3.1.2) may have a pole-type solution at the finite point if and only if the denominator in equation (3.3.1) has a zero at this point $q^2 = -m_g^2$ (in Euclidean signature),

$$-m_g^2 - m_g^2\Pi(-m_g^2;\xi) + \Delta^2 c(\xi) = 0, \tag{3.3.2}$$

where $m_g^2 \equiv m_g^2(\xi)$ is the effective gluon mass. The previous equation is a transcendental equation for its determination. Evidently, the number of its solutions is not fixed, *a priori*. Excluding the mass gap, one obtains that the denominator in the full gluon propagator becomes

$$q^2 + q^2\Pi\left(q^2;\xi\right) + \Delta^2 c\left(\xi\right) = q^2 + m_g^2 + q^2\Pi\left(q^2;\xi\right) + m_g^2\Pi\left(-m_g^2;\xi\right). \tag{3.3.3}$$

Let us now expand $\Pi(q^2;\xi)$ in a Taylor series near m_g^2:

$$\Pi\left(q^2;\xi\right) = \Pi\left(-m_g^2;\xi\right) + \left(q^2 + m_g^2\right)\Pi'\left(-m_g^2;\xi\right) + \mathcal{O}\left(\left(q^2 + m_g^2\right)^2\right). \tag{3.3.4}$$

Substituting this expansion into the previous relation and after doing some tedious algebra, one obtains

$$q^2 + m_g^2 + q^2\Pi\left(q^2;\xi\right) + m_g^2\Pi\left(-m_g^2;\xi\right) \\ = \left(q^2 + m_g^2\right)\left[1 + \Pi\left(-m_g^2;\xi\right) - m_g^2\Pi'\left(-m_g^2;\xi\right)\right]\left[1 + \Pi^R\left(q^2;\xi\right)\right], \tag{3.3.5}$$

where $\Pi^R(q^2;\xi) = 0$ at $q^2 = -m_g^2$ and it is regular at small q^2 values; otherwise it remains arbitrary.

The full gluon propagator (3.1.2) thus now looks

$$D_{\mu\nu}\left(q;m_g^2\right) = iT_{\mu\nu}(q)\frac{Z_3\left(m_g^2\right)}{\left(q^2+m_g^2\right)\left[1+\Pi^R\left(q^2;m_g^2\right)\right]} + i\xi L_{\mu\nu}(q)\frac{1}{q^2}, \tag{3.3.6}$$

where, for future purpose, in the invariant function $\Pi^R\left(q^2;m_g^2\right)$ instead of the gauge-fixing parameter ξ we introduced the dependence on the gluon effective mass squared m_g^2, which depends on ξ itself. The gluon propagator's renormalization constant is

$$Z_3\left(m_g^2\right) = \frac{1}{1 + \Pi\left(-m_g^2;\xi\right) - m_g^2\Pi'\left(-m_g^2;\xi\right)}. \tag{3.3.7}$$

In the formal limit $\Delta^2 = 0$, an effective gluon mass is also zero, $m_g^2(\xi) = 0$, as follows from equation (3.3.2). So an effective gluon mass is the non-perturbative effect. At the same time, it cannot be interpreted as the physical gluon mass, since it remains an explicitly gauge-dependent quantity (at least at this stage). In other words, we were unable to renormalize it along with the gluon propagator (3.3.6). In the formal $\Delta^2 = m_g^2(\xi) = 0$ limit the gluon propagator's renormalization constant (3.3.7) becomes the standard one [2, 3], namely

$$Z_3(0) = \frac{1}{1 + \Pi^s(0;\xi)}. \tag{3.3.8}$$

It is interesting to note that equation (3.3.2) has a second solution in the formal $\Delta^2 = 0$ limit. In this case an effective gluon mass remains finite, but $1 + \Pi\left(-m_g^2;\xi\right) = 0$. So a scale responsible for the non-perturbative dynamics is not determined by an effective gluon mass itself, but by this condition. Its interpretation from the physical point of view is not clear.

The general subtraction procedure described above can be directly applied to the massive solution (3.3.6). So it becomes

$$D_{\mu\nu}\left(q;m_g^2\right) = D_{\mu\nu}^{TNP}\left(q;m_g^2\right) + D_{\mu\nu}^{PT}\left(q\right), \tag{3.3.9}$$

where

$$D_{\mu\nu}^{TNP}\left(q;m_g^2\right) = iT_{\mu\nu}\left(q\right)\left[\frac{Z_3\left(m_g^2\right)}{\left(q^2+m_g^2\right)\left[1+\Pi^R\left(q^2;m_g^2\right)\right]} - \frac{Z_3\left(0\right)}{q^2\left[1+\Pi^R\left(q^2;0\right)\right]}\right] \tag{3.3.10}$$

and

$$D_{\mu\nu}^{PT}\left(q\right) = i\left[T_{\mu\nu}\left(q\right)\frac{Z_3\left(0\right)}{\left[1+\Pi^R\left(q^2;0\right)\right]} + \xi L_{\mu\nu}\left(q\right)\right]\frac{1}{q^2}. \tag{3.3.11}$$

The superscript 'TNP' stands for the truly non-perturbative gluon propagator. It differs from the intrinsically non-perturbative gluon propagator by the presence of the perturbation theory infrared singularity $1/q^2$ only, though in the formal perturbation theory $\Delta^2 = m_g^2(\xi) = 0$ limit it vanishes as well. In accordance with our prescription, we should finally replace the full gluon propagator (3.3.6) as follows: $D_{\mu\nu}(q;m_g^2) \to D_{\mu\nu}^{TNP}(q;m_g^2)$, where the latter is explicitly given in equation (3.3.10).

It is difficult to use the massive solution (3.3.6) for the solution of the color confinement problem, since it is smooth in the $q^2 \to 0$ limit. However, its existence shows the general possibility for a vector particles to acquire masses dynamically — without the so-called Higgs mechanism [44], which requires the existence of not yet fully discovered Higgs particle[1]. The above-mentioned possibility is due only to the internal dynamics and symmetries of the corresponding gauge theory. As a subject to discussion, let us point out that Higgs, being the fundamental scalar, may be the mass scale parameter, which has no any internal degrees of freedom, by definition. So, Higgs can be the mass scale parameter of the Standard Model in the same way as discussed here the Jaffe–Witten mass gap is for QCD. The Fermi constant of the electroweak model, having the dimensions of $[M]^{-2}$, apparently, is too weak indeed to generate any mass, especially such big masses of vector bosons.

[1]While we were finishing this book, CERN has been announced a possible discovery of a boson-like particle at around the theoretically-proposed Higgs mass with almost 5σ statistics. This might be the missing Higgs, but it has not been clarified yet.

3.4 Conclusions

The structure of the full gluon propagator in the presence of the regularized mass gap has been firmly established. We have shown explicitly that in its presence at least two independent and different types of formal solutions for the regularized full gluon propagator exist. No truncations/approximations/assumptions are made in order to show the existence of these general types of exact solutions. Also, our approach, in general, and the above-mentioned solutions, in particular, is gauge-invariant, since no special gauge has been chosen. Let us emphasize that before the renormalization program is performed the gauge invariance should be understood in this sense only.

In the presence of the mass gap the gluons may acquire an effective gluon masses, depending on the gauge choice (the so-called massive solution (3.3.6)), but a gauge-fixing parameter remains arbitrary — a gauge is not fixed by hand. Its relation to the solution of the color confinement problem is not clear, even after the renormalization program is performed. The general singular solution (3.2.10)–(3.2.13) for the full gluon propagator depends explicitly on the mass gap. It is always severely singular in the $q^2 \to 0$ limit, so the gluons remain massless, and this does not depend on the gauge choice.

The existence of some other solution(s) for the full gluon propagator should not be excluded *a priori*. Let us remind that the gluon Schwinger–Dyson equation (3.1.1) is a highly non-linear one, so the number of independent solutions is not fixed. Any concrete solutions obtained by the analytical approach based on the system of dynamical equations of motion are to be particular cases of the formal general solutions established here. Within our approach they are subject to the different truncations approximations/assumptions and the concrete gauge choice imposed on the invariant function $\Pi(q^2; D)$ in equation (3.1.1) (see, for example [17, 45] and recent papers [46–48] and references therein). It is worth emphasizing that any deviation of the full gluon propagator from the free one always assumes the presence of some mass scale parameter. However, it should be introduced without violating Slavnov–Taylor identity as it has been done in our approach. Let us also point out that the gluon propagator can be finite and contains the mass scale parameter [49–53], however it apparently cannot be interpreted as the gluon effective mass.

The intrinsically non-perturbative gluon propagator (3.2.16) is interesting for confinement, but the two important problems remain to be solved.

The first problem is how to perform the renormalization program for the regularized mass gap $\Delta^2 \equiv \Delta^2(\lambda, \alpha, \xi, g^2)$, and to see whether the mass gap survives it or not. The second problem is how to treat correctly severe infrared singularities $(q^2)^{-2-k}$, $k = 0, 1, 2, 3, ...$ inevitably present in this solution (let us remind that the signature is Euclidean). Both problems will be addressed and solved in the next Chapter.

3.A Appendix: The dimensional regularization method in the perturbation theory

As emphasized in our book, the mass gap is closely related to the skeleton tadpole term. Also, it has been explicitly shown that there is no such regularization scheme, preserving or not gauge invariance, in which the transversality condition for the full gluon self-energy could be satisfied unless the constant skeleton tadpole term (2.9.1), namely

$$\Pi^t_{\rho\sigma}(D) \equiv \Pi_t(D) \equiv \Delta^2_t(D) = g^2 \int \frac{id^4 q_1}{(2\pi)^4} T^0_4 D(q_1), \tag{3.A.1}$$

is to be disregarded from the very beginning, thus, put formally zero everywhere. Here T^0_4 is the four-gluon point-like vertex, and g^2 is the dimensionless coupling constant squared. We omit the tensor and color indices in this integral, as unimportant for further discussion. It is nothing else but the quadratically divergent constant in the perturbation theory, so it is assumed to be regularized. The mass gap does not survive in the perturbation theory $q^2 \to \infty$ limit, however, in this theory there are still problems with such kind of integrals.

In the perturbation theory, when the first non-trivial approximation for the full gluon propagator $D = D(q)$ is the free one $D_0 = D_0(q)$, the constant tadpole term is to be simply discarded, thus, to be put formally zero within the dimensional regularization method based on Ref. [29], so that $\Pi^t_{\rho\sigma}(D_0) = \delta_{\rho\sigma}\Delta^2_t(D_0) = 0$. However, even in the dimensional regularization method this is not an exact result, but rather an embarrassing prescription, as pointed out in Ref. [4]. To show explicitly that there are still problems, as mentioned above, it is instructive to substitute the first iteration of the gluon equation of motion into the previous expression (3.A.1). Symbolically it looks like

$$\begin{aligned} D(q) &= D_0(q) + D_0(q) i\Pi(q; D) D(q) \\ &= D_0(q) + D_0(q) i\Pi(q; D_0) D_0(q) + \dots \\ &= D_0(q) + D^{(1)}(q) + \dots , \end{aligned} \tag{3.A.2}$$

where we omit all the indices and put $D_0 \equiv D^{(0)}$.

Doing so, one obtains

$$
\begin{aligned}
\Pi_t(D) &= \Pi_t(D_0 + D^{(1)} + \dots) \\
&= \Pi_t(D_0) + \Pi_t(D^{(1)}) + \dots \\
&= \Pi_t(D_0) + g^2 \int \frac{id^4 q_1}{(2\pi)^4} T_4^0 [D_0(q_1)]^2 i\Pi(q_1; D_0) + \dots \\
&= \Pi_t(D_0) + \Pi_t(D_0) g^2 \int \frac{i^2 d^4 q_1}{(2\pi)^4} T_4^0 [D_0(q_1)]^2 \\
&\quad + g^2 \int \frac{i^2 d^4 q_1}{(2\pi)^4} T_4^0 [D_0(q_1)]^2 \Pi^s(q_1; D_0) + \dots .
\end{aligned}
\tag{3.A.3}
$$

Here we introduce the subtraction as follows: $\Pi^s(q_1; D_0) = \Pi(q_1; D_0) - \Pi(0; D_0)$, and put $\Pi(0; D_0) = \Pi_t(D_0)$, for simplicity, when the mass gap is to be reduced to the tadpole term itself. In the fourth line of (3.A.3) the first integral is not only ultraviolet divergent but infrared singular as well. If we now omit the first term in accordance with the above-mentioned prescription, the product of this integral and the tadpole term $\Pi_t(D_0)$ remains, nevertheless, undetermined. Moreover, the structure of the next integral is much more complicated than in the divergent constant integral $\Pi_t(D_0)$ in equation (3.A.1). All this reflects the general problem that such kinds of massless integrals

$$
\int \frac{\mathrm{d}^d q}{(2\pi)^d} \frac{q_{\mu_1} \dots q_{\mu_p}}{(q^2)^n}
\tag{3.A.4}
$$

are ill defined, since there is no dimension where they are meaningful. They are either infrared singular or ultraviolet divergent, depending on the relation between the numbers d, p and n [4]. This prescription clearly shows that the dimensional regularization method, preserving gauge invariance, nevertheless by itself is not sufficient to provide us insights into the correct treatment of the power-type infrared singularities shown in equation (3.A.4), which will be discussed in Appendix 3.B. Thus, one concludes that the tadpole term (3.A.1) $\Delta_t^2(D) \equiv \Delta_t^2(\lambda, \alpha; D)$ is, in general, not zero.

However, in the perturbation theory we can adhere to the prescription that such massless tadpole integrals can be discarded in the dimensional regularization method [4, 29]. As mentioned above, we have already shown that the mass gap, in general, and the skeleton tadpole term, in particular, can be neglected in the perturbation theory, indeed (not depending on whether λ, α are to be introduced within the regularization scheme preserving gauge invariance or not). In what follows we will show how precisely the dimensional regularization method [29] should be correctly implemented into the distribution theory in Ref. [40] in order to control the power-type severe infrared singularities, which may appear not only in the perturbation theory series, but mainly in the intrinsically non-perturbative QCD as well.

3.B Appendix: The dimensional regularization method in the distribution theory

In general, all the Green's functions in QCD are distributions. This is especially true for the non-perturbative infrared singularities of the full gluon propagator due to the self-interaction of massless gluons in the QCD vacuum. They present a rather broad and important class of functions with algebraic singularities, thus, functions with non-summable singularities at isolated points [40] (at zero in our case). Roughly speaking, this means that all relations involving distributions should be considered under corresponding integrals, taking into account the smoothness properties of the corresponding space of test functions. Let us note in advance that the space in which our generalized functions are continuous linear functionals is K, which is the space of infinitely differentiable functions having compact support, thus, they are zero outside some finite region — different for each differentiable function [40].

Let us consider the positively definite, $P > 0$ quadratic Euclidean form

$$P(q) = q_0^2 + q_1^2 + q_2^2 + ... + q_{n-1}^2 = q^2, \tag{3.B.1}$$

where n is the number of the components. The distribution $P^\lambda(q)$, where λ is, in general, an arbitrary complex number, is defined as

$$(P^\lambda, \varphi) = \int_{P>0} P^\lambda(q)\varphi(q) \ \mathrm{d}^d q, \tag{3.B.2}$$

where $\varphi(q)$ is the above-mentioned test function. At $\mathrm{Re}(\lambda) \geq 0$ this integral is convergent and is an analytic function of λ. Analytical continuation to the region $\mathrm{Re}(\lambda) < 0$ shows that it has a simple pole at points [40]

$$\lambda = -\frac{n}{2} - k, \qquad k = 0, 1, 2, 3... \tag{3.B.3}$$

In order to actually define the skeleton loop integrals in the deep infrared domain, which appear in the system of the Schwinger – Dyson equations, it is necessary to introduce the infrared regularization parameter ϵ, defined as $d = n + 2\epsilon$, $\epsilon \to 0^+$ within the dimensional regularization method [29], where d is the dimension of the loop integral based on equation (3.B.2). As a result, all the Green's functions and bare parameters should be regularized with respect to ϵ which should be set to zero at the end of the computations. The structure of the severe infrared singularities is then determined (when n is even number) as follows [40]:

$$(q^2)^\lambda = \frac{C_{-1}^{(k)}}{\lambda + (d/2) + k} + \text{finite terms}, \tag{3.B.4}$$

where the residue is

$$C_{-1}^{(k)} = \frac{\pi^{n/2}}{2^{2k} k! \Gamma((n/2) + k)} \cdot L^k \delta^n(q) \tag{3.B.5}$$

with

$$L = \frac{\partial^2}{\partial q_0^2} + \frac{\partial^2}{\partial q_1^2} + \cdots + \frac{\partial^2}{\partial q_{n-1}^2}. \tag{3.B.6}$$

Thus the regularization of the severe infrared singularities (3.B.4), on account of (3.B.5), is nothing but the whole expansion in the corresponding powers of ϵ and not the separate term(s). Let us underline its most remarkable feature. The order of singularity does not depend on λ, n and k. In terms of the infrared regularization parameter ϵ it is always a simple pole $1/\epsilon$. This means that all power terms in equation (3.B.4) will have the same singularity, thus,

$$(q^2)^{-\frac{n}{2}-k} = \frac{1}{\epsilon} C_{-1}^{(k)} + \text{finite terms}, \quad \text{while} \quad \epsilon \to 0^+, \tag{3.B.7}$$

where we can put $d = n$ now, after introducing this expansion. By 'finite terms' here and everywhere a number of necessary subtractions under corresponding integrals is understood [40]. However, the residue at a pole will be drastically changed from one power singularity to another. This means different solutions to the whole system of the dynamical equations for different set of numbers λ and k. Different solutions mean, in their turn, different vacua. In this picture different vacua are to be labeled by the two independent numbers: the exponent λ and k. At a given number of $d(=n)$ the exponent λ is always negative being an integer if $d(=n)$ is an even number or fractional if $d(=n)$ is an odd number. The number k is always an integer and positive and precisely it determines the corresponding residue at a simple pole, see equation (3.B.7). It would not be surprising if these numbers were somehow related to the nontrivial topology of the QCD vacuum in any dimensions.

It is worth emphasizing that the structure of severe infrared singularities in Euclidean space is much simpler than in Minkowski space, where kinematical (unphysical) singularities due to the light cone also exist, as in Refs. [2, 40, 238]. In this connection let us remind that in Euclidean metrics $q^2 = 0$ implies $q_i = 0$ and *vice versa*, while in Minkowski metrics this is not so. In this case it is rather difficult to untangle them correctly from the dynamical singularities, the only ones which are important for the calculation of any physical observable. Also, the consideration is much more complicated in the configuration space [40]. That is why we always prefer to work in the momentum space — where propagators do not depend explicitly on the number of dimensions — with Euclidean signature. We also prefer to work in covariant gauges in order to avoid peculiarities of the non-covariant gauges as in Refs. [2, 239, 240], for example, how to untangle the gauge pole from the dynamical one.

In principle, none of the regularization schemes — how to introduce the infrared regularization parameter in order to parameterize severe infrared divergences and thus to put them under control — should be introduced by hand. First of all, it should be well defined. Secondly, it should be compatible with the distribution theory [40]. The dimensional regularization method [29] is well defined, and here we have shown how it should be introduced into the distribution theory — complemented by the number of subtractions, if necessary. Though the so-called $\pm i\epsilon$ regularization is formally equivalent to the regularization used in our book, nevertheless, it is rather inconvenient for practical use. Especially this is true for the gauge-field propagators, which are substantially modified due to the response of the vacuum — the $\pm i\epsilon$ prescription is designated for and is applicable only to the theories with the perturbation theory vacua, indeed [65, 66]. Other regularization schemes are also available, for example, such as analytical regularization used in Ref. [34] or the so-called Speer's regularization [241]. However, they should be compatible with the distribution theory as emphasized above. Anyway, the regularization is not important but the distribution theory itself. Just this theory provides an adequate mathematical framework for the correct treatment of all the Green's functions in QCD (apparently, for the first time the distribution nature of the Green's functions in quantum field theory has been recognized and used in Ref. [242]).

The regularization of the non-perturbative infrared singularities in QCD is determined by the Laurent expansion (3.B.7) at $n = 4$ as follows:

$$(q^2)^{-2-k} = \frac{1}{\epsilon} a(k) \left[\delta^4(q)\right]^{(k)} + \text{finite terms} = \frac{1}{\epsilon}\left[a(k)[\delta^4(q)]^{(k)} + \mathcal{O}_k(\epsilon)\right], \tag{3.B.8}$$

where $\epsilon \to 0^+$ and $a(k) = \pi^2/2^{2k}\, k!\, \Gamma(2+k)$ is a finite constant depending only on k and $\left[\delta^4(q)\right]^{(k)}$ represents the kth derivative of the δ-function. see equations (3.B.4) and (3.B.5). We point out that after introducing this expansion everywhere one can fix the number of dimensions, thus, put $d = n = 4$ for QCD without any further problems. Indeed there will be no other severe infrared singularities with respect to ϵ as it goes to zero, but those explicitly shown in this expansion. Effectively, ϵ is a gauge invariant version of the initial α. However, even if there is no dependence on α in the mass gap after the renormalization procedure the dependence on ϵ will again appear, see Section 4.3 and remarks above.

Let us note in advance that, while the initial expansions (4.2.1)–(4.2.2) is the Laurent expansion in the inverse powers of the gluon momentum squared, the regularization expansion (3.B.8) is the Laurent expansion in powers of ϵ. This means that its regular part is as follows:

$$\text{finite terms} = \left(q^2\right)_-^{-2-k} + \epsilon \left(q^2\right)_-^{-2-k} \ln q^2 + \mathcal{O}\left(\epsilon^2\right), \qquad (3.B.9)$$

where for the definition of the functional $(q^2)_-^{-2-k}$ see Ref. [40]. These terms, however, play no role in the infrared multiplicative renormalization program which will be discussed in Chapter 4. The dimensionally regularized expansion (3.B.8) takes place only in four-dimensional QCD with Euclidean signature. In other dimensions and/or Minkowski signature it is much more complicated as pointed out above. As it follows from this expansion any power-type non-perturbative infrared singularity, including the simplest one at $k = 0$, scales as $1/\epsilon$ as it goes to zero. Just this plays a crucial role in the infrared renormalization of the theory within our approach. Evidently, such a kind of the dimensionally regularized expansion (3.B.8) does not exist for the perturbation theory infrared singularity, which is as much singular as $(q^2)^{-1}$ only.

In summary, first we have emphasized the distribution nature of the non-perturbative infrared singularities. Secondly, we have explicitly shown how the dimensional regularization method should be correctly and in a gauge invariant way implemented into the distribution theory. This makes it possible to put severe infrared singularities under firm mathematical control.

Problems

Problem 3.1. *Derive the relation given by equation* (3.2.3).

Problem 3.2. *Derive the decomposition* (3.2.7).

Problem 3.3. *Derive the expression* (3.3.5).

Problem 3.4. *Re-write equation* (3.2.1) *as equation* (2.7.1). *Perform the direct non-linear iteration procedure in order to get expression* (3.2.7). *For guidance see Ref.* [37].

Chapter 4

Renormalization of the Mass Gap

4.1 Introduction

In previous Chapters the mass gap responsible for the non-perturbative dynamics in QCD has been introduced. It is mainly generated by the non-linear interaction of massless gluon modes. However, through the tadpole term the most important role in its generation belongs to the point-like four-gluon vertex. We have explicitly shown that QCD is the color gauge invariant theory at non-zero mass gap as well. All this allows one to establish the structure of the full gluon propagator in the explicit presence of the mass gap. In this case, the two independent general types of the formal solutions for the regularized full gluon propagator have been found. No truncations/approximations/assumptions, as well as no special gauge choice are made for the regularized skeleton loop integrals, contributing to the full gluon self-energy. The so-called massive solution, which leads to an effective gluon mass has been established. The general singular solution for the full gluon propagator depends explicitly on the mass gap. It is always severely singular in the $q^2 \to 0$ limit, so the gluons remain massless, and this does not depend on the gauge choice. However, we have argued that only its intrinsically non-perturbative part — which is free of all the types of perturbative contributions — is interesting for confinement. In this way it becomes automatically transversal in a gauge invariant way and excludes the free gluons from the theory. This is important for correct understanding of the color confinement mechanism. The intrinsically non-perturbative gluon propagator is to be used for the numerical calculations of physical observables, processes, etc. in low-energy QCD from first principles. In this Chapter we present the solution for the two important problems mentioned at the end of the previous Chapter.

4.2 The intrinsically non-perturbative gluon propagator

The full gluon propagator, which is relevant for the description of the true QCD dynamics, the so-called intrinsically non-perturbative gluon propagator (3.2.16)–(3.2.17) derived in the previous chapter is as follows:

$$D_{\mu\nu}^{INP}(q) = iT_{\mu\nu}(q)\, d^{INP}(q^2)\frac{1}{q^2} = iT_{\mu\nu}(q)\frac{\Delta^2}{(q^2)^2}L(q^2), \tag{4.2.1}$$

and

$$L(q^2) \equiv L(q^2;\Delta^2) = \sum_{k=0}^{\infty}\left(\frac{\Delta^2}{q^2}\right)^k \Phi_k = \sum_{k=0}^{\infty}\left(\frac{\Delta^2}{q^2}\right)^k \sum_{m=0}^{\infty}\Phi_{km}, \tag{4.2.2}$$

where $\Delta^2 \equiv \Delta^2(\lambda,\alpha,\xi,g^2)$ and the residues $\Phi_k \equiv \Phi_k(\lambda,\alpha,\xi,g^2)$ and thus $\Phi_{km} \equiv \Phi_{km}(\lambda,\alpha,\xi,g^2)$ as well.

The intrinsically non-perturbative gluon propagator (4.2.1) depends only on the transversal degrees of freedom of gauge bosons. Its functional dependence is uniquely fixed up to the expressions for the residues Φ_k, and it is valid in the whole energy/momentum range. It explicitly depends on the mass gap, so that when it formally goes to zero (the formal perturbation theory $\Delta^2 = 0$ limit) this solution vanishes. Also, it is characterized by the explicit presence of severe (i.e., non-perturbative) infrared singularities $(q^2)^{-2-k}$, $k = 0, 1, 2, 3, \ldots$ only. The sum over m indicates that an infinite number of the corresponding contributions invokes each severe infrared singularity labeled by k. Let us remind that these contributions come from the skeleton loop integrals, contributing to $c(d)$ and $\Pi(q^2; d)$ in the $q^2 \to 0$ limit in equation (3.2.1). Apart from the structure $(\Delta^2/(q^2)^2$ it is nothing but the corresponding Laurent expansion. Apparently, this expansion in integer powers of (Δ^2/q^2) is of cluster-type expansion discussed in [11]. Indeed, expansion is done in powers of the mass gap (accompanied by the corresponding inverse powers of the gluon momentum squared) and not in an elementary way in powers of the coupling constant. The dependence on it in this expansion is rather non-implicitly complicated. The summation over k and m shows that the intrinsically non-perturbative gluon propagator (4.2.1)–(4.2.2) incorporates all severe infrared singularities possible in QCD at non-zero mass gap. So all other full vertices with gluon momenta involved can be considered as free of such kind of singularities — they become regular functions of their arguments. This solution is completely free of all the types of the perturbative contributions (contaminations), and thus is exactly and uniquely separated from the perturbation theory gluon propagator. Due to the regular dependence on the mass gap, it dominates over the free gluon propagator in the deep infrared ($q^2 \to 0$) limit and it is suppressed in the deep ultraviolet ($q^2 \to \infty$) limit. Let us remind that the perturbative gluon propagator $D^{PT}_{\mu\nu}(q)$ and its free counterpart $D^0_{\mu\nu}(q)$ behave like $1/q^2$ in the whole momentum range — apart from the log corrections in the $q^2 \to \infty$ limit for $D^{PT}_{\mu\nu}(q)$ due to the asymptotic freedom. For simplicity, in the rest of this chapter we will omit the superscript 'INP'. We will restore it once it is necessary for clarity of our discussion.

4.3 Confining gluon propagator

One of the remarkable features of the solution (4.2.1)–(4.2.2) is that its asymptotic at infinity ($q^2 \to \infty$) is to be determined by its $\Delta^2/(q^2)^2$ structure only, since all other terms in its expansion are suppressed in this limit. It is well known that such a behavior at infinity is not dangerous for the renormalizability of QCD [2–5]. However, the regularized mass gap itself is of the non-perturbative origin. So its renormalization in order to get a finite result in the $\lambda \to \infty$ limit cannot be done in the standard way. Another main problem, closely connected to the mass gap, is its violent structure in the deep infrared region ($q^2 \to 0$).

However, a new surprising feature of this solution is that its structure at zero ($q^2 \to 0$) can be again determined by its $\Delta^2/(q^2)^2$ term only. To show this explicitly, let us begin with the theorem from the theory of functions of complex variable [54], which is extremely useful for the explanation of the behavior of our solution (4.2.1)–(4.2.2) in the deep infrared $q^2 \to 0$ limit.

The function $L(q^2)$ is defined by its Laurent expansion (4.2.2), and thus it has an isolated essentially singular point at $q^2 = 0$. Its behavior in the neighborhood of this point is regulated by the Weierstrass–Sokhatsky–Casorati theorem (see Appendix 4.A), which reveals that

$$\lim_{n \to \infty} L(q_n^2) = Z, \quad q_n^2 \to 0, \tag{4.3.1}$$

where Z is any complex number, and q_n^2 is a sequence of points $q_1^2, q_2^2, \dots, q_n^2, \dots$ along which q^2 goes to zero, and for which the above-displayed limit always exists. Of course, Z remains arbitrary (it depends on the chosen sequence of points q_n^2) but, in general, it depends on the same set of parameters as the residues thus, $Z \equiv Z(\lambda, \alpha, \xi, g^2)$. This theorem thus allows one to replace the Laurent expansion $L(q^2)$ by Z when $q^2 \to 0$ independently from all other test functions in the corresponding skeleton loop integrands,

$$L\left(q^2\right) \to Z\left(\lambda, \alpha, \xi, g^2\right), \quad q^2 \to 0. \tag{4.3.2}$$

Our consideration in this Chapter up to this point is necessarily formal, since the mass gap remains unrenormalized. So far it has been only regularized, as $\Delta^2 \equiv \Delta^2\left(\lambda, \alpha, \xi, g^2\right)$. The renormalization of the mass gap can proceed as follows. Due to the above-formulated Weierstrass – Sokhatsky – Casorati theorem, the full gluon propagator (4.2.1) effectively becomes

$$D_{\mu\nu}(q) = iT_{\mu\nu}(q)\ \frac{1}{\left(q^2\right)^2}\ Z\left(\lambda, \alpha, \xi, g^2\right)\ \Delta^2\left(\lambda, \alpha, \xi, g^2\right), \quad q^2 \to 0, \tag{4.3.3}$$

so just the $\Delta^2\left(q^2\right)^{-2}$ structure of the full gluon propagator (4.2.1) is all that matters, indeed. Let us now define the renormalized ('R') mass gap as follows:

$$\Delta_R^2 = Z\left(\lambda, \alpha, \xi, g^2\right)\ \Delta^2\left(\lambda, \alpha, \xi, g^2\right), \tag{4.3.4}$$

so that we consider $Z\left(\lambda, \alpha, \xi, g^2\right)$ as the multiplicative renormalization constant for the mass gap, and Δ_R^2 is the physical mass gap within our approach. Precisely this quantity should be positive, finite, gauge-independent, etc., and it should exist when $\lambda \to \infty$ and $\alpha \to 0$. Due to the Weierstrass – Sokhatsky – Casorati theorem, we can always choose such $Z = Z_1 \cdot Z_2 \cdot \ldots$ in order to satisfy all the necessary requirements (each Z_n will depend on its own sequence of points along which $q^2 \to 0$, the so-called subsequences).

Thus, the full gluon propagator relevant for the intrinsically non-perturbative QCD dynamics becomes

$$D_{\mu\nu}(q; \Delta_R^2) = iT_{\mu\nu}(q)\ \frac{\Delta_R^2}{(q^2)^2}, \tag{4.3.5}$$

where from now on we introduce into the gluon propagator the explicit dependence on the mass gap, for future purposes. It is possible to say that because of the Weierstrass – Sokhatsky – Casorati theorem there always exists such a sequence of points along which $q^2 \to 0$. So our main result can be formulated as follows:

The intrinsically non-perturbative gluon propagator (4.2.1)–(4.2.2) *effectively converges to the function* (4.3.5) *in the whole q^2-momentum plane (containing both points $q^2 = 0$ and $q^2 = \infty$) after the renormalization of the mass gap is performed.*

But the region of small q^2 is of special interest. In this region the confinement dynamics begin to play a dominant role. However, it is worth emphasizing that the solution (4.3.5) is not the infrared asymptotic ($q^2 \to 0$) of the initial expansion (4.2.1)–(4.2.2), since $(q^2)^{-2}$ term is not a leading one in this limit, but it is a leading one just in the opposite $q^2 \to \infty$ limit.

The severe infrared singularity $(q^2)^{-2}$, of which the only one is present in the relevant gluon propagator (4.3.5), is the first non-perturbative infrared singularity possible in four-dimensional QCD. A special regularization expansion is to be used in order to deal with it, since the standard methods fail in this case. As mentioned above, it should be correctly treated within the distribution theory [40] complemented by the dimensional regularization method [29]. If q^2 is an independent skeleton loop variable, then the corresponding dimensional regularization of this singularity is given by the expansion, namely

$$\left(q^2\right)^{-2} = \frac{1}{\epsilon}\left[a(0)\delta^4(q) + \mathcal{O}_0(\epsilon)\right], \quad \epsilon \to 0^+, \tag{4.3.6}$$

and $a(0) = \pi^2$ (see expansion (3.B.8) at $k = 0$ in Appendix 3.B). Due to the $\delta^4(q)$ function in the residue of this expansion, all the test functions which appear under corresponding skeleton loop integrals should finally be replaced by their expression at $q = 0$. So the dimensionally regularized expansion for the gluon propagator (4.3.5) becomes

$$D_{\mu\nu}\left(q; \Delta_R^2\right) = iT_{\mu\nu}\left(q\right)\Delta_R^2 \cdot \frac{1}{\epsilon}\left[a(0)\delta^4(q) + \mathcal{O}(\epsilon)\right], \quad \epsilon \to 0^+, \tag{4.3.7}$$

where we put $\mathcal{O}_0(\epsilon) = \mathcal{O}(\epsilon)$, for simplicity. As emphasized in Appendix 3.B, in the presence of the infrared regularization parameter ϵ all the Green's functions and parameters become, in general, dependent on it. The only way to remove the pole $1/\epsilon$ from the relevant gluon propagator (4.3.7) is to define the infrared renormalized mass gap as follows:

$$\Delta_R^2 = X\left(\epsilon\right)\bar{\Delta}_R^2 = \epsilon \cdot \bar{\Delta}_R^2, \quad \epsilon \to 0^+, \tag{4.3.8}$$

where $X(\epsilon) = \epsilon$ is the infrared multiplicative renormalization constant for the mass gap. Let us remind that contrary to Δ_R^2 its infrared renormalized counterpart $\bar{\Delta}_R^2$ exists as $\epsilon \to 0^+$. In both expressions for the mass gap the dependence on ϵ is assumed but not shown explicitly.

It is well known that the expression (4.3.5) leads to the linear rising potential between heavy quarks [55] also seen by lattice QCD [56, 57]. Nevertheless, we conventionally call it the 'confining gluon propagator' instead of the 'confining potential', because it is a distribution. Just like the generalized function it will be used for the derivation of the system of equations determining the confining quark propagator (for preliminary consideration see [58] and references therein).

4.4 The renormalized running effective charge

It is instructive to find explicitly the corresponding β-function. From equation (4.3.5) it follows that the corresponding Lorentz structure, which is nothing but the corresponding effective running charge, in terms of the renormalized mass gap is

$$d\left(q^2;\Delta_R^2\right) \equiv \alpha\left(q^2;\Delta_R^2\right) = \frac{\Delta_R^2}{q^2}, \tag{4.4.1}$$

and this does not depend on whether the gluon momentum is an independent loop variable or not. Then from the renormalization group equation for the renormalized effective charge, which determines the β-function,

$$q^2 \frac{d\alpha\left(q^2;\Delta_R^2\right)}{dq^2} = \beta\left(\alpha\left(q^2;\Delta_R^2\right)\right), \tag{4.4.2}$$

it simply follows that

$$\beta\left(\alpha\left(q^2;\Delta_R^2\right)\right) = -\alpha\left(q^2;\Delta_R^2\right) = -\frac{\Delta_R^2}{q^2}. \tag{4.4.3}$$

Thus, one can conclude that the corresponding β-function as a function of its argument is always in the domain of attraction, and thus negative. So it has no infrared stable fixed point indeed as it is required for the confining theory [2]. Just these expressions for the β-function and the running effective charge should be used for the calculation of the non-perturbative quantities in the Yang–Mills theory, such as the gluon condensate, the gluon part of the bag constant [59–61], etc. (see also Part II of this book). In phenomenology we need to know for these purposes the ratio

$$\frac{\beta\left(q^2\right)}{\alpha\left(q^2\right)} \equiv \frac{\beta\left(\alpha\left(q^2;\Delta_R^2\right)\right)}{\alpha\left(q^2;\Delta_R^2\right)} = -1, \tag{4.4.4}$$

so within our approach it does not depend on q^2 at all and don't forget that the replacement $\alpha(q^2) \to \alpha^{INP}(q^2)$ and $\beta(q^2) \to \beta^{INP}(q^2)$ are assumed.

Concluding, let us note that in the renormalized effective charge (4.4.1) the perturbation theory infrared singularity $(1/q^2)$ is only present. So there is no need for the additional infrared renormalization of the mass gap. However, if it will be multiplied by the additional powers of $(q^2)^{-2-k}$ $k = 0, 1, 2, 3, \dots$, then this product should be treated as the full gluon propagator itself. The only difference between them becomes the unimportant tensor structure $T_{\mu\nu}(q)$.

4.5 The general criterion of gluon confinement

We are now in the position to analytically formulate the general criterion of gluon confinement. The infrared renormalization of the mass gap (4.3.8) automatically infrared renormalizes the relevant gluon propagator (4.3.5) as follows:

$$D_{\mu\nu}(q) = \epsilon \cdot iT_{\mu\nu}(q)\bar{\Delta}_R^2(q^2)^{-2}, \qquad \epsilon \to 0^+. \tag{4.5.1}$$

This is nothing but the dimensionally renormalized (within the distribution theory complemented by the dimensional renormalization method) expression for the relevant gluon propagator in the intrinsically non-perturbative QCD[1]. There is no doubt left that the Weierstrass – Sokhatsky – Casorati theorem underlines the importance of the simplest non-perturbative infrared singularity $1/(q^2)^2$ possible in four-dimensional QCD, while all others are suppressed in the deep infrared ($q^2 \to 0$) region due to this theorem.

Due to the distribution nature of the non-perturbative infrared singularity $(q^2)^{-2}$, the two principally different cases should be separately considered (see Appendix 3.B for the details).

(i) *The gluon momentum q is an independent skeleton loop variable.*

As repeatedly emphasized above, then the initial non-perturbative infrared singularity $(q^2)^{-2}$ should be regularized with the help of the expansion (4.3.6). So one finally arrives at the dimensionally regularized expansion in the $\epsilon \to 0^+$ limit (4.5.1), which finally becomes

$$D_{\mu\nu}\left(q; \bar{\Delta}_R^2\right) = iT_{\mu\nu}\left(q\right)\bar{\Delta}_R^2 a\left(0\right)\delta^4\left(q\right) + \mathcal{O}\left(\epsilon\right), \quad \epsilon \to 0^+. \tag{4.5.2}$$

Evidently, after performing the renormalization program (i.e., going to the infrared renormalized quantities), the terms of the order $\mathcal{O}(\epsilon)$ can be omitted from the consideration.

[1] Let us emphasize that this is the general expression for the renormalized gluon propagator, since it does not depend on whether the gluon momentum is an independent skeleton loop variable or not.

Let us note in advance that beyond the one-loop skeleton integrals the analysis should be done in a more sophisticated way. Otherwise in the multi-loop skeleton diagrams containing gluon propagators the appearance of the product of at least two δ-functions at the same point is possible. However this product is not defined in the distribution theory [40]. So in the multi-loop skeleton diagrams instead of the δ-functions in the residues their derivatives have to appear, see again [40] and Appendix 3.B. They should be treated in the sense of the distribution theory. Fortunately, the infrared renormalization of the theory is not undermined, since a pole in ϵ is always a simple pole $1/\epsilon$ for each independent skeleton loop variable (see the dimensionally regularized expansion (3.B.8)). Following Ref. [38], the general multiplicative infrared renormalization program of the intrinsically non-perturbative gluon propagator (4.2.1)–(4.2.2) can be done.

(ii) *The gluon momentum q is not a skeleton loop variable.*
Then the first necessary and second sufficient conditions for gluon confinement are:

(a) The first necessary condition of gluon confinement is the absence of the dressed gluons in the physical spectrum (by the dressed gluons we mean the gluons whose propagation is described by the full gluon propagator). If the gluon momentum q is not a loop variable then it is the external momentum. Then the initial $(q^2)^{-2}$ non-perturbative infrared singularity cannot be treated as the distribution. Then the regularization expansion (4.3.6) is not the case to be used. The function $(q^2)^{-2}$ is the standard one, and the relevant gluon propagator (4.5.2) vanishes as ϵ goes to zero, i.e.,

$$D_{\mu\nu}(q) = \epsilon \cdot iT_{\mu\nu}(q)\bar{\Delta}_R^2(q^2)^{-2} \sim \epsilon, \quad \epsilon \to 0^+. \tag{4.5.3}$$

It is worth emphasizing that the final $\epsilon \to 0^+$ limit can be taken only after expressing all the Green's functions, parameters and the mass gap in terms of their infrared renormalized counterparts because they, by definition, exist in this limit. This behavior is gauge-invariant, does not depend on any truncations/approximations/assumptions, etc., and thus it is a general one. It prevents the transversal dressed gluons from appearing in asymptotic states, so color dressed gluons can never be isolated.

(b) The second sufficient condition of gluon confinement is the absence of the free gluons in the corresponding theory. Just such a theory has been formulated in our previous works [36–38]. We call it the intrinsically non-perturbative QCD. Let us remind that its full gluon propagator (4.2.1) has no free gluon propagator limit when the interaction is to be switched off ($\Delta^2 = 0$). We consider both the suppression of the colored dressed gluons at large distances and the absence of the free colored gluons in the theory as the exact criterion of gluon confinement (for its initial formulation see Refs. [39, 62]). Concluding, let us note that in [63] the gluons depending on the loop variable are called virtual ones, while the gluons not depending on the loop variable are called actual ones. Of course, this separation is only convention because of the self-interaction of massless gluons, but, nevertheless, it is useful to understand quark dynamics in the vacuum of QCD.

A few general remarks are in order. In the functional space we should distinguish between the purely transversal large gluon fields, which decrease more slowly than $1/r$ (in fact, they increase at large distances r, at least linearly), and the gluon fields of an arbitrary covariant gauge, which decrease at least as $1/r$ [64, 65]. In Ref. [66] the so-called *Gribov's dilemma* has been formulated as follows: "the solution of the confinement problem lies not in the understanding of the interaction of 'large gluon fields' but instead in the understanding of how the QCD dynamics can be arranged as to prevent the non-Abelian fields from growing real big". In other words, if indeed the non-linear interaction of 'large gluon fields' is responsible for color confinement, then why can't we see them?

Within our approach formulated in momentum space the above mentioned separation between increasing and decreasing gluon fields corresponds to the purely transversal severely infrared singular virtual gluon fields and gluon fields of arbitrary gauge, respectively. It is worth reminding that this exact and uniquely defined separation is to be traced back to the principal difference between severely (i.e., non-perturbative) and perturbative infrared singularities. The purely transversal severely infrared singular *virtual* gluon fields are indeed prevented from growing really big within our approach. The $\delta^4(q)$ infrared singularity in equation (4.5.2) is much weaker than the initial one in equation (4.5.1). Moreover, it makes the whole expression for the corresponding correlation function/observable finite. At the same time, the purely transversal severely infrared singular *actual* gluon fields are indeed suppressed in equation (4.5.3), that is why we do not see them. This explains the *Gribov's dilemma*.

4.6 The general criterion of quark confinement

It is instructive to formulate here the quark confinement criterion in advance as well. It consists of the two independent conditions.

(i) The first necessary condition, formulated at the fundamental, microscopic quark-gluon level, is the absence of pole-type singularities in the quark Green's function at any gauge on the real axis at some finite point in the complex momentum plane,

$$S(p) \neq \frac{Z_2}{\hat{p} - m_{ph}}, \tag{4.6.1}$$

where Z_2 is the standard quark wave function renormalization constant, while m_{ph} is the mass to which a physical meaning could be assigned. In other words, the quark always remains an off-mass-shell object. Such an interpretation of quark confinement comes apparently from the Gribov's approach to quark confinement [65] and Preparata's massive quark model in which external quark legs were approximated by entire functions [67]. A quark propagator may or may not be an entire function, but in any case the pole of the first order — like the electron propagator has in QED — should disappear as for example in Refs. [17, 18, 68–70] and references therein.

(ii) The second sufficient condition, formulated at the hadronic, macroscopic level, is the existence of the discrete spectrum only in bound-states, in order to prevent quarks from appearing in asymptotic states. This condition comes apparently from the 't Hooft's model for two-dimensional QCD with large N_c limit [71] (see also Refs. [34, 68]). This definition of the quark confinement in the momentum space is gauge invariant and flavor independent. It is valid for all types of quarks (even light or heavy) and thus it is a general one. The Wilson criterion of quark confinement formulated in the configuration space-area law [72, 73] is relevant only for heavy quarks, as well as a linear rising potential between static (heavy) quarks [55], also carried out by lattice QCD [56, 57].

In the quark–gluon plasma [74, 75] most of the bound-states will be dissolved, so the second sufficient condition does not work any more. However, the first necessary condition remains always valid. By increasing temperature or density there is no way to put quarks and gluons on the mass-shell. The colored gluons can indeed propagate from one hadron to the next in the thermal state with many overlapping hadrons [76, 77]. But, nevertheless, they are not free, thus $q^2 \neq 0$. So what is called as de-confinement phase transition in the quark–gluon plasma is, in fact, de-hadronization phase transition. De-confinement at the fundamental quark–gluon level is about the liberation of the colored objects from the vacuum and not from the bound-states. In the QCD ground-state there are many colored objects such as quarks, gluons, instantons and maybe something else. Since color confinement is absolute and permanent, none of these colored objects can appear in physical spectrum, and thus de-confinement phase transition does not exist, in principle.

4.7 The general criterion of dynamical/spontaneous breakdown of chiral symmetry

In close connection with quark confinement is another important nonperturbative effect of dynamical or, equivalently, spontaneous breakdown of chiral symmetry. We have already mentioned in Chapter 1 that in the Lagrangian of QCD the chiral $SU(N_f) \times SU(N_f)$ flavor symmetry is explicitly broken by the current quark mass term. So in the chiral limit this symmetry can be broken only by the nontrivial dynamics of QCD related to quark confinement and other way around [78–82]. The general criterion of dynamical breakdown of chiral symmetry also consists of the two conditions.

(i) The first necessary condition: The general decomposition of the inverse of the quark Green's function is $S^{-1}(p) = i\left[\hat{p}A\left(p^2\right) + B\left(p^2\right)\right]$ and its free counterpart is $S_0^{-1}(p) = i\left[\hat{p} + m_0\right]$ with m_0 being the current (bare) quark mass (Euclidean signature). If the confining solution for the quark propagator is regular at zero $p^2 = 0$ then the effective quark mass can be defined as

$S^{-1}(0) = iB(0) = im_{eff}$ in complete analogy with the current quark mass $S_0^{-1}(0) = im_0$ — of course, any finite renormalization point is a possible choice. However, it seems that the true confining solution has to be regular at zero as written in Ref. [58] and references therein. On the other hand, from the Schwinger–Dyson equation for the quark propagator at zero it follows that $m_{eff} = m_0 + m_d(\lambda)$, where

$$m_d\left(\lambda\right) \sim \int d^4q \Gamma_\mu\left(q\right) S\left(q\right) \gamma_\nu D_{\mu\nu}\left(q\right), \tag{4.7.1}$$

and $\Gamma_\mu\left(q\right)$ is the full quark–gluon vertex and $D_{\mu\nu}\left(q\right)$ is the gluon propagator. So in the chiral limit $m_0 = 0$, one obtains: primarily, if $m_{eff} = m_d\left(\lambda\right) \neq 0$, then it is the *chiral symmetry violating* solution; secondly if $m_{eff} = m_d(\lambda) = 0$, then it is the *chiral symmetry preserving* solution.

Chiral symmetry breakdown of the quark propagator on general grounds can be formulated as follows:

$$\left\{S^{-1}\left(p\right), \gamma_5\right\}_+ = i\gamma_5 2B\left(p^2\right) \neq 0, \tag{4.7.2}$$

while for the free quark propagator it is zero in the chiral limit $m_0 = 0$. Let us emphasize that the first necessary condition of dynamical breakdown of chiral symmetry (4.7.2) is gauge invariant, thus it takes place at any gauge, and the measure of this breakdown is always twice dynamically generated, running quark mass (4.7.1). In this connection it is worth reminding that the true confining quark propagator cannot be gauge dependent. Inequality (4.7.2) is the first necessary condition of dynamical breakdown of chiral symmetry at the fundamental quark level.

(ii) The second sufficient condition: However, there exists one more condition of chiral symmetry breakdown at the phenomenological level as well. The measure of the phenomenological breakdown of chiral symmetry is the non-zero chiral quark condensate. It is defined as follows:

$$\left\langle 0 \left|\bar{q}q\right| 0\right\rangle_0 \sim \int \mathrm{d}^4p \mathrm{Tr}\, S\left(p\right) \sim -\int \mathrm{d}^4p \frac{B\left(p^2\right)}{p^2 A^2\left(p^2\right) + B^2\left(p^2\right)}, \tag{4.7.3}$$

where subscript '0' means that it should be calculated in the chiral limit. So one can conclude that if there is no dynamical breakdown

of chiral symmetry at the fundamental quark level, thus $B(p^2) = 0$, then the chiral condensate is also always zero, but not *vice versa*. The dynamically generated quark mass $B(p^2) \neq 0$, while the chiral quark condensate can be still zero, in principle. In other words, if the chiral quark condensate becomes zero, this does not automatically means that dynamical quark mass also becomes zero.
Lattice QCD shows that in the quark–gluon plasma the chiral quark condensate drastically goes down at some characteristic temperature $T_c \approx 180$ MeV, indeed [84, 197]. However, as mentioned above, the thermal fluctuations encompass states with many overlapping hadrons [76, 77]. So propagating from one hadron to the next, gluons will only increase the contributions to the self-energy (4.7.1) of the corresponding quarks. Using Mandelstam's [63] terminology, quarks gain contributions to their effective masses not only from the virtual gluons but from the actual gluons via three- and four-gluon vertices as well. Above T_c all the quarks which have been liberated from the hadrons and which have not yet (i.e., still remaining in hadrons) become heavier. The chiral condensate decreases but the quark–gluon plasma is saturated by the massive quarks especially above T_c, as it has been predicted by the quark coalescence model [85, 86]. The thermal chiral condensate vanishes but not thermally excited dynamically generated quark masses. We can conclude that there is no chiral symmetry restoration phase transition in the quark–gluon plasma at the fundamental quark and gluon level. It is again only the de-hadronization phase transition.

Quarks degrees of freedom always work against pure gluon degrees of freedom, since quark loops are coming with overall negative sign. Near T_c there are a lot of massive and massless gluonic excitations and fluctuations, as well as massive quarks and their clusters of different origin. Furthermore, all this are encompassed by the states of overlapping hadrons, etc. Thus, the nature of the phase transition across T_c in the quark–gluon plasma, in principle, should not be so sharp as in the pure gluon plasma, where it is of the first order (see Part II of this book). In fact, it is a smooth crossover by lattice QCD in Ref. [87] and references therein.

4.8 Physical limits

We introduced the renormalized mass gap in the relation (4.3.4), defined its existence when the dimensionless ultraviolet regulating parameter λ goes to infinity. However, nothing was said about the behavior of the coupling constant squared g^2 in this limit. In principle, it may also depend on λ, becoming thus the so-called running effective charge $g^2 \sim \alpha_s \equiv \alpha_s(\lambda)$. In the general composition (4.3.4)

$$Z(\lambda, \alpha_s(\lambda)) \Delta^2(\lambda, \alpha_s(\lambda)), \tag{4.8.1}$$

all the possible types of the effective charge behavior in the $\lambda \to \infty$ limit should be considered independently from each other[2].

(i) If $\alpha_s(\lambda) \to \infty$ as $\lambda \to \infty$, then one recovers the strong coupling regime. If this limit for the composition (4.8.1) exists then it can be defined as the renormalized mass gap (4.3.4),

$$Z(\lambda, \alpha_s(\lambda)) \Delta^2(\lambda, \alpha_s(\lambda)) = \Delta_R^2, \text{ with } \quad \lambda \to \infty \ \& \ \alpha_s(\lambda) \to \infty. \tag{4.8.2}$$

Apparently, only this mass gap can be identified/related with/to the Jaffe–Witten mass gap discussed in Ref. [11].

(ii) If $\alpha_s(\lambda) \to c$ as $\lambda \to \infty$, where c is a finite constant, then it can be put as unity, without loss of generality. This means that the effective charge becomes unity, and this is only possible for the free gluon propagator. But the free gluon propagator contains none of the mass scale parameters, so in fact

$$Z(\lambda, \alpha_s(\lambda)) \Delta^2(\lambda, \alpha_s(\lambda)) = 0 \text{ with } \quad \lambda \to \infty \ \& \ \alpha_s(\lambda) \to 1. \tag{4.8.3}$$

(iii) If $\alpha_s(\lambda) \to 0$ as $\lambda \to \infty$, then one recovers the weak coupling regime. Evidently, in this finite limit, the value of the mass scale parameter can be identified as Λ^2_{QCD},

$$Z(\lambda, \alpha_s(\lambda)) \Delta^2(\lambda, \alpha_s(\lambda)) = \Lambda^2_{QCD}, \text{ with } \quad \lambda \to \infty \ \& \ \alpha_s(\lambda) \to 0. \tag{4.8.4}$$

Of course, in each case there are different Zs. Let us underline that the above-mentioned finite limits always exist indeed due to the Weirstrass-Sokhatsky-Casorati theorem cited above in Section 4.3 (see also Appendix 4.A below). There is no doubt that the regularized mass gap may provide

[2]The dependence on other parameters is not shown in this Section, as it is unimportant for further discussion.

the existence of the two different physical mass scale parameters after the renormalization program is performed. Though these two physical parameters show up explicitly at different regimes, nevertheless, numerically they may not be very different, indeed. Our mass gap Δ_R^2 determines the power-type deviation of the full gluon propagator from the free one in the $q^2 \to 0$ limit. The region of small q^2 is interesting for all the non-perturbative effects in QCD. This once more emphasizes the close link between the behavior of QCD at large distances and its intrinsically non-perturbative dynamics. At the same time, the asymptotic QCD scale parameter Λ_{QCD}^2 determines much weaker logarithmic deviation of the full gluon propagator from the free one in the $q^2 \to \infty$ limit. Then an interesting question arises within our approach. How exactly does the regularized mass gap provide the appearance of Λ_{QCD}^2 under the perturbative logarithms? The problem is that in the full gluon propagator the regularized mass gap contribution is linearly suppressed in comparison with the logarithmical divergent term in the perturbation theory $q^2 \to \infty$ limit.

4.9 Asymptotic freedom and the mass gap

In the perturbation theory the effective charge is given by the expression $d^{PT}(q^2) = \left[1 + \Pi\left(q^2; d^{PT}\right)\right]^{-1}$. Within this, the invariant function $\Pi\left(q^2; d^{PT}\right)$ can be only logarithmically divergent in the $q^2 \to \infty$ limit at any d, in particular at $d = d^{PT}$. So putting for further convenience $d^{PT}(q^2) = \alpha_s\left(q^2; \Lambda^2\right)/\alpha_s(\lambda)$ in this relation, one obtains

$$\alpha_s\left(q^2; \Lambda^2\right) = \frac{\alpha_s(\lambda)}{1 + b\alpha_s(\lambda)\ln\left(q^2/\Lambda^2\right)}, \tag{4.9.1}$$

and b is the standard color group factor. This expression represents the summation of the so-called main perturbative logarithms in powers of $\alpha_s(\lambda)$. However, in this book it has been obtained without solution of the renormalization group equations. Evidently, nothing should depend on Λ — and hence on λ too — when they go to infinity in order to recover the finite effective charge in this limit. To show explicitly that this finite limit exists, let us formally write

$$\Lambda^2 = f(\lambda)\Delta^2(\lambda, \alpha_s(\lambda)), \tag{4.9.2}$$

which is always valid, since $f(\lambda)$ is, in general, an arbitrary dimensionless function. In this connection, let us again remind that in order to get the expression (4.9.1) from the full effective charge $d\left(q^2\right) =$

$\left[1+\Pi\left(q^2;d\right)+c\left(d\right)\left(\Delta^2/q^2\right)\right]^{-1}$ the mass gap contribution Δ^2/q^2 is only asymptotically suppressed in the $q^2 \to \infty$ limit. In other words, we distinguish between the asymptotic suppression of the mass gap contribution Δ^2/q^2 and the formal $\Delta^2 = 0$ limit. So the mass gap $\Delta^2 = \Delta^2\left(\lambda, \alpha_s\left(\lambda\right)\right)$ itself here is not put identically zero, and hence the relation (4.9.2) makes sense. On account of the relation (4.8.4), it becomes

$$\Lambda^2 = f\left(\lambda\right) Z^{-1}\left(\lambda, \alpha_s\left(\lambda\right)\right) \Lambda^2_{QCD}, \text{ with } \lambda \to \infty \ \&\ \alpha_s\left(\lambda\right) \to 0. \quad (4.9.3)$$

Substituting this into the expression (4.9.1) and doing some algebra, one obtains

$$\alpha_s\left(q^2\right) = \frac{\alpha_s}{1 + b\,\alpha_s \ln\left(q^2/\Lambda^2_{QCD}\right)}, \quad (4.9.4)$$

if and only if

$$\alpha_s = \frac{\alpha_s\left(\lambda\right)}{1 - b\,\alpha_s\left(\lambda\right)\ln\left(f/Z\right)}, \text{ with } \lambda \to \infty \ \&\ \alpha_s\left(\lambda\right) \to 0 \quad (4.9.5)$$

exists and is finite in the above shown limits. Here we introduce the shorthand notations $f \equiv f\left(\lambda\right)$ and $Z \equiv Z\left(\lambda, \alpha_s\left(\lambda\right)\right)$, for simplicity. Evidently, the finite α_s can be identified with the fine structure constant of the strong interactions, calculated at some fixed scale. It is worth emphasizing that the existence and finiteness of α_s is due to the ratio, f/Z. Indeed, from equation (4.9.5) it follows

$$\ln\left(f/Z\right) = \frac{\alpha_s - \alpha_s\left(\lambda\right)}{\alpha_s b \alpha_s\left(\lambda\right)} \to \frac{1}{b\,\alpha_s\left(\lambda\right)}, \text{ with } \lambda \to \infty \ \&\ \alpha_s\left(\lambda\right) \to 0, \quad (4.9.6)$$

which means that $\left(f/Z\right) = \exp\left(1/b\,\alpha_s\left(\lambda\right)\right)$ in the above shown limits. Substituting this into the relation (4.9.3) it becomes

$$\lim_{(\Lambda,\lambda)\to\infty} \Lambda^2 \exp\left[-\frac{1}{b\,\alpha_s\left(\lambda\right)}\right] = \Lambda^2_{QCD}, \text{ with } \alpha_s\left(\lambda\right) \to 0, \quad (4.9.7)$$

which is the finite limit of the renormalization group equations solution.

At very large q^2 from equation (4.9.4) one recovers

$$\alpha_s(q^2) = \frac{1}{b \ln\left(q^2/\Lambda^2_{QCD}\right)}, \quad (4.9.8)$$

which is the asymptotic freedom famous formula if $b > 0$. In QCD with three colors and six flavors this is so, indeed. For the pure Yang–Mills fields $b = 11/4\pi > 0$ always. Let us underline that in the expressions (4.9.4) and (4.9.8) q^2 is always big enough, so it cannot go below Λ^2_{QCD}. We have shown explicitly the asymptotic freedom behavior of QCD at short distances ($q^2 \to \infty$), not using the renormalization group equations and their solutions — we need no expansion in powers of the coupling constant for the corresponding β-function [2, 3, 5, 88]. The regularized mass gap is suppressed in the $q^2 \to \infty$ limit, as it has been mentioned above. Based on equation (4.8.4) and our consideration in this subsection it follows, nevertheless, that the regularized mass gap in the $\lambda \to \infty$ limit provides the existence of the asymptotic QCD scale parameter Λ^2_{QCD} as well.

There is no relation between the renormalized mass gap Δ^2_R (4.8.2) and the asymptotic scale parameter Λ^2_{QCD} (4.8.4), since they show up explicitly at different regimes. They are different scales, indeed, responsible for different non-perturbative and non-trivial perturbative dynamics in QCD, though numerically they may not be very different, as underlined above. However, originally they have been generated in the region of small q^2. In Ref. [88] it has been noticed that being numerically a few hundred MeV only, Λ^2_{QCD} cannot survive in the $q^2 \to \infty$ limit. So none of the finite mass scale parameters can be determined by the perturbative QCD. They should come from the region of small q^2, being thus non-perturbative by origin and surviving the renormalization program (the removal of λ in the $\lambda \to \infty$ limit), as has just been demonstrated above.

Concluding, all this can be a manifestation that "the problems encountered in perturbation theory are not mere mathematical artifacts but rather signify deep properties of the full theory" [89]. The message that we are trying to convey is that the intrinsically non-perturbative dynamical structure of the full gluon propagator indicates the existence of its nontrivial perturbative structure and *vice versa*.

4.A Appendix: The Weierstrass – Sokhatsky – Casorati theorem

One of the main theorems in the theory of functions of complex variable is the Weierstrass – Sokhatsky – Casorati theorem [54]. It describes the behavior of meromorphic functions near essential singularities.

Weierstrass – Sokhatsky – Casorati theorem: *If z_0 is an essential singularity of the function $f(z)$, then for any complex number Z there exists the sequence of points $z_k \to z_0$, such that*

$$\lim_{k\to\infty} f(z_k) = Z. \tag{4.A.1}$$

So this theorem tells us that the behavior of the function $f(z)$ near its essential singularity z_0 is not determined, i.e, in fact, it remains arbitrary. It depends on the chosen sequence of points z_k along which z goes to zero, that's $Z \equiv Z(z_k)$.

Let us consider a classical example taken from Ref. [54]. The function

$$f(z) = e^{1/z} = \sum_{n=0}^{\infty} \frac{1}{n!}\frac{1}{z^n} \tag{4.A.2}$$

has the above shown Laurent series about the essential singularity at $z = 0$, thus, this Laurent expansion converges to the function $f(z)$ everywhere apart from the point $z = 0$. At the same time, due to the Weierstrass – Sokhatsky – Casorati theorem the behavior of the function $f(z)$ near $z = 0$, and hence of the Laurent expansion itself, depends on the chosen sequence of points z_k along which $z \to 0$. So let us proceed as follows:

(i) If one chooses $z_k = 1/k$, with $k = 1, 2, 3, ...$, then

$$\lim_{k\to\infty} f(z_k) = \lim_{k\to\infty} e^k = \infty. \tag{4.A.3}$$

(ii) If one chooses $z_k = -1/k$, with $k = 1, 2, 3, ...$, then

$$\lim_{k\to\infty} f(z_k) = \lim_{k\to\infty} e^{-k} = 0. \tag{4.A.4}$$

(iii) If one chooses $z_k = 1/\ln A + 2k\pi i$, with $k = 0, 1, 2, 3, ...$, then

$$\lim_{k\to\infty} f(z_k) = \lim_{k\to\infty} e^{\ln A + 2k\pi i} = A, \tag{4.A.5}$$

where A is some finite constant.

Many other examples can be found in Ref. [54] and in other text-books on the theory of functions of complex variable.

Concluding, let us make one important point perfectly clear. For example, the function

$$f(z) = \frac{z}{\sqrt{1+z^2}} = \sum_{n=0}^{\infty} (-1)^n \frac{(2n)!}{2^{2n}(n!)^2} \frac{1}{z^{2n}} \tag{4.A.6}$$

has no singularity at zero at all, while its Laurent expansion always has an essential singularity at $z = 0$, by definition. The equality (4.A.6) means that this Laurent expansion converges to the above shown function $f(z)$ in the ring which excludes zero point. Its region of convergence is $1 < |z| < \infty$, while the behavior of any Laurent expansion near its essential singularity is always governed by the Weierstrass–Sokhatsky–Casorati theorem, in particular the Laurent expansion shown in the right-hand-side of equation (4.A.6).

Another characteristic example is the function

$$f(z) = \frac{z}{z-1} = \sum_{n=0}^{\infty} \left(\frac{1}{z}\right)^n, \tag{4.A.7}$$

which is nothing but a geometric series. The Laurent expansion, shown in the right-hand-side of this equation, converges to the function $f(z)$ also in the region $1 < |z| < \infty$, while the behavior of the Laurent expansion itself at $z \to 0$ is again governed by the Weierstrass–Sokhatsky–Casorati theorem. The same is true for the Laurent expansion in equation (4.A.2). It converges to the function $\exp(1/z)$ in the whole complex plane apart from the point $z = 0$, while near this point its behavior is uncertain, as it was described above. The message we are trying to convey is that all the equalities containing the Laurent expansions should be treated carefully in accordance with the above-mentioned theorem, thus the region of convergence should be fixed clearly, otherwise the equality can be incorrectly understood.

Problems

Problem 4.1. *Perform the general multiplicative infrared renormalization program for the the intrinsically non-perturbative gluon propagator* (4.2.1)–(4.2.2). *For guidance see Appendix 3.B and Ref.* [38].

Problem 4.2. *Substituting equation* (4.4.1) *into equation* (4.7.1), *show that for the confining gluon propagator m_d is finite and non-zero, thus the chiral symmetry in the confining phase dynamically breaks down.*

Problem 4.3. *Derive relations given by equations* (4.9.4) *and* (4.9.5).

Problem 4.4. *Derive relations* (4.9.6) *and* (4.9.7).

Chapter 5

General Discussion

5.1 Discussion

The true QCD ground state is a very complicated confining medium, containing many types of gluon field configurations, components, ingredients and objects of different nature [2, 90–96]. Its dynamical and topological complexity means that its structure can be organized at both the quantum and classical levels. It is definitely contaminated by such gluon field excitations and fluctuations, which are of the perturbative origin, nature and magnitude. Moreover, it may contain such extra gluon field configurations, which cannot be detected as possible solutions to the QCD dynamical equations of motion, either quantum or classical, for example vortex-type [97, 98] or string ones [99]. The only well known classical component of the QCD ground state is the topologically nontrivial instanton–antiinstanton type of fluctuations of gluon fields, which are solutions to the Euclidean Yang – Mills classical equations of motion in the weak coupling regime [44, 100, 101]. However, they are by no means dominant but, nevertheless, playing a special role in the QCD vacuum. In our opinion their main task is to prevent quarks and gluons from freely propagating in the QCD vacuum. It seems to us that this role does not contradict their standard interpretation such as tunneling trajectories linking vacua with different topology [2, 44] (for more detailed qualitative discussion see Ref. [68]).

What we have shown in our investigation here is that the QCD vacuum is really beset with severe infrared singularities of dynamical origin due to the existence of the mass gap in QCD. Moreover, we have identified the main source of these singularities as the four-gluon vertex, which generates the tadpole term, and hence the mass gap itself. In other words, these singularities are always coming together with the mass gap and *vice versa*. They have been summarized (accumulated) into the intrinsically non-perturbative gluon propagator (4.2.1)–(4.2.2). There is no doubt that *the purely transversal severely infrared singular virtual gluon fields* play an important role in the dynamical structure of the QCD ground state, leading thus to the effect of zero momentum modes enhancement there [102, 166]. We have explicitly shown indeed that the low-frequency components of the transversal virtual gluon fields in the true vacuum should have larger amplitudes than those of a perturbative ('bare') vacuum [63]. "But it is to just this violent infrared behavior that we must look for the key to the low energy and large distance hadron phenomena. In particular, the absence of quarks and other colored objects can only be understood in terms of the infrared divergences in the self-energy of a color bearing objects" [107].

Evidently, such mechanism of confinement is nothing but the well-forgotten infrared slavery one, which can be equivalently referred to as a strong coupling regime [2, 34]. Indeed, at the very beginning of QCD it was expressed as a general idea [1, 34, 63, 104–108] that the quantum excitations of the infrared degrees of freedom, because of self-interaction of massless gluons in the QCD vacuum, made it only possible to understand confinement, dynamical (spontaneous) breakdown of chiral symmetry and other non-perturbative effects. In other words, the importance of the deep infrared structure of the QCD vacuum has been emphasized as well as its relevance to the above-mentioned non-perturbative effects and the other way around. This development was stopped by the well-known wrong opinion that severe infrared singularities cannot be put under control. Here we have explicitly shown that the adequate mathematical theory for quantum Yang–Mills gauge theory is the distribution theory (the theory of generalized functions) [40], complemented by the dimensional regularization method [29]. Together with the theory of functions of complex variable [54] they provide a correct treatment of these severe infrared singularities without any problems (see Appendix 3.B as well). Thus, we come back to the old idea but on a new basis that is why it becomes new ('new is well-forgotten old'). In other words, we put the infrared slavery mechanism of color confinement on a firm mathematical ground. For example, the correct treatment of these severe infrared singularities and the renormalization of the mass gap made it possible to analytically formulate the gluon confinement criterion in a gauge invariant way for the first time.

In general we are working in the momentum space, where we deal with the purely transversal severely infrared singular gluon fields. Discussing the relevant field configurations, we will always mean the configuration space. All the possible low-frequency components (large scale amplitudes) of the QCD vacuum are purely transversal virtual fields. These are important for the dynamical formation of such gluon field configurations which are responsible for gluon confinement and other non-perturbative effects within our approach to low-energy QCD. In other words, collective motion of all the purely transversal *virtual* gluon field configurations with low-frequency components/large scale amplitudes is solely responsible for gluon confinement. At the same time, the amplitudes of all the purely transversal *actual* gluon field configurations are totally suppressed (no transversal gluons at large distances). Let us note that at this stage, it is difficult to identify which type of gauge field configurations can finally be formed by the purely transversal severely infrared singular gluon fields in the QCD ground state — to identify relevant field configurations: chromomagnetic, self-dual, stochastic, etc.

The confining gluon propagator (4.3.5) in the different approximations for the full gluon propagator in the different gauges has been earlier obtained and investigated, as well as used like an ansatz, in many papers, see for example [18, 63, 68, 109–118] and references therein. From the very beginning we have used the confining gluon propagator (4.3.5) as an ansatz, mainly paying attention to its functional dependence [68]. And only becoming aware of the Jaffe – Witten theorem, we realized that the dependence on the mass scale parameter (the mass gap) is equally important, especially for the renormalization program. There exists also direct lattice evidence that the zero momentum modes are enhanced in the full gluon propagator [119–122]. A non-perturbative finite-size scaling technique was used in [123] to study the evolution of the running coupling in $SU(3)$ Yang – Mills lattice theory. By using the two-loop β-function it is shown to evolve according to perturbation theory at high energies, while at low energies it is shown to grow (see also [124]). Some classical models of the QCD vacuum also invoke a $(q^2)^{-2}$ behavior of the gluon fields in the infrared domain. For example, it appears in the QCD vacuum as a condensation of the color-magnetic monopoles (QCD vacuum is a chromomagnetic superconductor) proposed by Nambu, Mandelstam and 't Hooft and developed by Nair and Rosenzweig (see [99, 125] and references therein). In the classical mechanism of the confining medium [126] and in the effective theory for the QCD vacuum proposed in [127] and to derive the linearly rising potential between heavy quarks within the renormalization group flow equations approach [128, 129] this singularity is also required.

We have confirmed and thus revitalized all such kind of investigations, in which this behavior has been obtained as an infrared asymptotical solution to the gluon Schwinger – Dyson equation or has been used as a confining ansatz in different quantum and classical models and approaches to the QCD ground state. However, let us emphasize that our initial intrinsically non-perturbative gluon propagator (4.2.1)–(4.2.2), which is valid in the whole momentum range, is an exact result. It summarizes all the severe infrared singularities generated by the interactions of massless gluon modes, and surviving only due the existence of the mass gap in QCD. Moreover, the Weierstrass – Sokhatsky – Casorati theorem clearly shows that only the simplest of them $1/(q^2)^2$ survives the renormalization of the mass gap. So only the $\Delta_R^2/(q^2)^2$ structure is important, while all other terms in the intrinsically non-perturbative gluon propagator (4.2.1) are suppressed — though each next term in the expansion (4.2.2) is more singular in the infrared than the previous one. In other words, the whole cluster expansion (4.2.1)–(4.2.2) effectively converges to the function (4.3.5) in the whole momentum plane after the renormalization of the mass gap is performed. So it is neither the infrared asymptotic of the full gluon propagator nor the confining ansatz; it is an exact result, indeed.

The distribution nature of $(q^2)^{-2}$ infrared singularity has been also underlined in [130, 131], but without recognizing the important role of the infrared renormalization of their mass scale parameter μ^2 in order to remove a pole $1/\epsilon$. It inevitably appears after the correct implementation of the dimensional regularization method into the distribution theory, as it has been demonstrated in Appendix 3.B. The simple replacement $(q^2)^{-2}$ infrared singularity by the $\delta^4(q)$-function was a reason why they were unable to formulate the gluon confinement criterion. Such a trivial replacement could not get a future development, since it failed in multi-loop skeleton diagrams, leading to a possible multiplication of a few $\delta^4(q)$-functions at the same point. As emphasized above, such a product is ill defined in the distribution theory. And finally, the $\delta^4(q)$-type potential in the configuration space leads to the constant potential between heavy quarks, i.e., not to the linear rising one.

There exists an interesting analogy between two- and four-dimensional QCD, where $(q^2)^{-1}$ and $(q^2)^{-2}$ in the framework of the distribution theory are the simplest severe infrared singularities, respectively. Two-dimensional QCD confines quarks [32, 71, 132]. Though the QCD vacuum is a much more complicated confining medium than its two-dimensional model, nevertheless, the above-mentioned analogy is promising in order to guide how to confine not only heavy but light quarks as well to four-dimensional QCD ground state.

In the presence of the mass gap the coupling constant plays no role. This is also an evidence of the 'dimensional transmutation', $g^2 \to \Delta^2(\lambda, \alpha, \xi, g^2)$ [2, 133, 134], which occurs whenever a massless theory acquires masses dynamically. It is a general feature of spontaneous symmetry breaking in field theories. In our case, the color gauge symmetry is explicitly broken by the tadpole term at the level of the full gluon self-energy, while dynamically maintained by the mass gap at the level of the full gluon propagator. In the massive solution (3.3.8) the mass gap transforms further into the effective gluon mass, i.e., $g^2 \to \Delta^2(\lambda, \alpha, \xi, g^2) \to m_g(\xi)$, but it remains gauge-dependent even after the corresponding full gluon propagator is renormalized. Within the intrinsically non-perturbative QCD the mass gap transforms further into the physical mass gap, i.e., $g^2 \to \Delta^2(\lambda, \alpha, \xi, g^2) \to \Delta_R^2$, and the gluons remain massless in a gauge invariant way.

5.2 Subtractions

Let us continue our discussion by recalling that many important quantities in QCD, such as gluon and quark condensates, topological susceptibility, the bag constant, etc., are defined beyond perturbation theory only [59, 95, 135–137]. They are determined by such S-matrix elements (correlation functions) from which all types of the perturbation theory contributions should be, by definition, subtracted. In this book we have already advocated in previous chapters the necessity of the first subtraction at the fundamental (microscopic) gluon propagator level. At the hadronic (macroscopic) level, the second step is to integrate out quark and gluon degrees of freedom. It aims at the subtractions of the perturbative parts in the corresponding integrals in term of which this or that correlation function is usually expressed. Making the subtractions at both levels of all the types of the perturbation theory contributions ('contaminations'), all the correlations functions will be expressed in terms of the finite integrals. It is worth emphasizing that by subtracting an infinity from another infinity, we render the theory free of the perturbation theory ultraviolet divergences — in the $\lambda \to \infty$ limit. This is a standard procedure of the renormalization program in order to get a finite result for the correlation functions. On the other hand, let us remind that at the fundamental gluon propagator level the subtraction is nothing but adding zero, as it was explicitly demonstrated in equation (2.8.4). The same is true for the subtractions at the hadronic/phenomenological level as well (see Part II of this book). So both types of the advocated subtractions at the microscopic and macroscopic levels do not violate Lorentz invariance from the very beginning.

Let us emphasize that such type of subtractions are inevitable also for the sake of self-consistency. In low-energy QCD there exist relations between different correlation functions, for example the famous Witten–Veneziano and Gell-Mann–Oakes–Renner formulae. The former [138–140] relates the pion decay constant and the mass of the η' meson to the topological susceptibility. The latter [135, 141] relates the chiral quark condensate to the pion decay constant and its mass. Defining the topological susceptibility and the quark condensate by the subtraction of all types of the perturbation theory contributions, it would not be self-consistent to retain them in the correlation functions determining the pion decay constant and in the expressions for the pion and η' meson masses.

There exists also an additional but very serious argument in favor of the inevitability of the above-discussed subtractions of the perturbation theory contributions at all levels in order to fix the gauge of the intrinsically non-perturbative QCD. In Refs. [64, 65] the quantization problem of non-Abelian gauge theories using the functional integral representation of the generating functional for non-Abelian gauge fields has been investigated. It has been explicitly shown that the standard Fadeev–Popov prescription fails to fix the gauge uniquely and therefore should be modified — it is not enough to eliminate arbitrary degrees of freedom from the theory. In other words, there is an ambiguity in the gauge-fixing of non-Abelian gauge fields — the so-called Gribov ambiguity (uncertainty), which results in Gribov copies and *vice versa.* To resolve this problem Gribov has explicitly demonstrated that the above-mentioned modification reduces simply to an additional limitation on the integration range in the functional space of non-Abelian gauge fields, which consists in integrating only over the fields for which the Fadeev–Popov determinant is positive (introducing thus the so-called Gribov horizon in the functional space). The intrinsically non-perturbative QCD is a manifestly gauge-invariant at the fundamental gluon level. All problems with the gauge fixing discovered by Gribov in the functional space should be attributed to the perturbative QCD within our approach. Subtracting further the perturbative contaminations at the hadronic (microscopic) level, we thus will make the intrinsically non-perturbative QCD free of the gauge-fixing ambiguity in the momentum space. Finally this will lead to the existence of something like Gribov horizon in functional space but in the momentum space. We would like to emphasize that the general proposal to subtract all the types of the perturbation theory contaminations at all levels is our solution to the problem of Gribov copies, which otherwise will plague the dynamics of any essentially non-linear gauge systems [66].

A few remarks on the subtraction of the perturbation theory contributions are in order. In lattice QCD [142] such an equivalent procedure also exists. In order to prepare an ensemble of lattice configurations for the calculation of any non-perturbative quantity or to investigate some non-perturbative phenomena, the excitations and fluctuations of gluon fields of the perturbative origin and magnitude should be washed out from the vacuum. This goal is usually to be achieved by using 'Perfect Actions' [143], 'cooling' [144], 'cycling' [145], etc. (see also [90, 91, 93–95] and references therein). Evidently, in lattice QCD this is very similar to our method in continuous QCD in order to proceed to the intrinsically non-perturbative QCD. However, there exists also a principal difference. In lattice QCD, at final stage one should go to the continuum limit first and only then to the infinite volume limit and not *vice versa*, while the chiral limit should be taken last, if necessary (however, care is needed because of the chiral log problem) [146]. These are parts of the renormalization program in order to remove the ultraviolet and infrared cutoffs. So, all numerical lattice results for any non-perturbative quantity, somehow become again inevitably contaminated by finite (but might be very small) perturbative contributions — and thus will be plagued by the above-mentioned Gribov uncertainties. There is no way to escape this fundamental difficulty. The criterion of the quark confinement — area law — derived by Wilson in lattice gauge theory [72] becomes inadequate for the continuous theory [147], since the Lorentz invariance is restored but confinement is lost. Contrary to lattice QCD, the intrinsically non-perturbative QCD becomes free of the perturbative contaminations forever, even after the removal of the ultraviolet cutoff in the $\lambda \to \infty$ limit.

In QCD sum rules [135, 136, 148] the corresponding subtraction should be also done in order to calculate, for example such non-perturbative quantity as the gluon condensate. While in our approach we should subtract finally the intrinsically non-perturbative part of the full gluon propagator integrated out over the perturbative region (see Part II of this book), in QCD sum rules one needs to subtract the perturbative solution of the full gluon propagator integrated out over the deep infrared region, where it certainly fails (see discussion given by Shifman in [91]). The necessity of the subtraction of the perturbative part of the effective coupling constant (integrated out) in order to correctly calculate the gluon condensate by analytic method has been explicitly shown in Ref. [149] as well.

All these physical arguments in favor of the subtraction of all the types of the perturbation theory contributions at all levels are completely justified by the distribution theory, which requires the integration over the finite region (in momentum space) in the integrals describing this or that physical quantity. This is due to the fact that all the Green's functions in QCD are continuous linear functionals in K, that's the space of functions, having compact support, thus, they are zero outside some finite region. See discussion in Appendix 3.B.

5.3 Conclusions

Let us denote the version of the mass gap $\bar{\Delta}_R^2$ which will appear in the S-matrix elements for the corresponding physical quantities/processes in low-energy QCD as Λ_{INP}^2 (in principle, they may be slightly different from each other, indeed). Then a symbolic relation between it and the initial mass gap $\Delta^2(\lambda, \alpha_s(\lambda))$ and Λ_{PT}^2 instead of Λ_{QCD}^2 (for reason see discussion below) could be written as follows:

$$\Lambda_{INP}^2 \xleftarrow[\infty \leftarrow \lambda]{\infty \leftarrow \alpha_s(\lambda)} \Delta^2(\lambda, \alpha_s(\lambda)) \xrightarrow[\lambda \to \infty]{\alpha_s(\lambda) \to 0} \Lambda_{PT}^2, \tag{5.3.1}$$

which summarizes our main results obtained in Part I of this book. QCD as a quantum gauge field theory, describing the interactions of never seen colored objects (gluons and quarks), cannot have the physical mass gap. In other words, this is a theory which describes the behavior of the colored objects in the vacuum. In QCD mass gap may only appear in the way described here, that's $\Delta^2(\lambda, \alpha_s(\lambda))$ in equation (5.3.1). Its full gluon propagator is given by system of equations (2.6.1)–(2.6.3). In order to become the theory of strong interactions it should undergo the two phase transitions: in the weak and strong coupling regimes. In the first case it becomes the perturbative, '*PT*' QCD which describes all the high-energy phenomena in the strong interactions from first principles: asymptotic freedom, scale violation, hard processes, etc. It has its own physical mass gap which we denote as Λ_{PT}^2 in equation (5.3.1). Its full gluon propagator is given by system of equations (2.6.4)–(2.6.6), while at large q^2 its effective charge becomes (4.9.4) with the above-advocated replacement, $\Lambda_{QCD}^2 \to \Lambda_{PT}^2$. In the second case it becomes the intrinsically non-perturbative (INP) QCD which describes all the low-energy phenomena in the strong interactions from first principles: those includes first of all color confinement, dynamical breakdown of chiral symmetry, bound-states, etc. It has its own physical mass gap, that's Λ^2_{INP} in equation (5.3.1) and its full gluon propagator is given by equation (4.3.5). At the final stage of the calculations in INP QCD the renormalized mass gap is replaced by Λ^2_{INP} thus, $\Delta^2_R \to \Lambda^2_{INT}$.

In this connection, a few things should be made perfectly clear. First of all, let us underline that the perturbative QCD and the intrinsically non-perturbative QCD are not effective theories. As pointed out above both theories are fundamental ones. Secondly, such a quantity as Λ^2_{QCD} does not exist at all, since QCD itself cannot have a physical limit, as repeatedly underlined above. In order to avoid any confusion the corresponding scale is better denoted as Λ^2_{PT}, since just the perturbative QCD is responsible for all the high-energy phenomena in the strong interactions. Thus similar to the relation (5.3.1), the following symbolic relation makes sense

$$\boxed{\text{INP QCD}} \Longleftarrow \boxed{\text{QCD}} \Longrightarrow \boxed{\text{PT QCD}} \tag{5.3.2}$$

so that at the fundamental (quark–gluon) level the perturbative QCD is asymptotically free, while the intrinsically non-perturbative QCD confines gluons, and we hope that it will confine quarks as well (for preliminary consideration see Ref. [58]).

A few years ago Jaffe and Witten formulated the theorem [11], which has been explicitly presented and discussed in Chapter 1, and which was the starting point of our investigations in [36–38]. Of course, to prove the existence of the Yang–Mills theory with compact simple gauge group G is still a formidable task. It is a mathematical, rather than a physical problem. However, from the Jaffe–Witten presentation of their theorem it clearly follows that their mass gap should be identified with our mass gap Λ^2_{INP}. At the same time, we have argued above that QCD itself cannot have a physical mass gap. It has a mass gap which is only regularized, $\Delta^2 \equiv \Delta^2(\lambda, \alpha_s(\lambda))$, and therefore there is no guarantee that it is positive. It cannot be related directly to any of physical quantities/processes. Let us also remind that QCD itself cannot confine free gluons. As actual theory of the strong interactions the two different faces of QCD come into the play: perturbative QCD for high-energy physics and intrinsically non-perturbative QCD for low-energy physics, which confines dressed gluons, while free gluons do not exist in this theory. The corresponding mass gaps have now physical meanings: they are finite, positive, gauge-invariant, etc.

Our basic results obtained in previous Chapters can be jointly formulated as follows:

Mass Gap Existence and Gluon Confinement: *If quantum Yang–Mills theory with compact simple gauge group $G = SU(3)$ exists on $\mathbb{R}^4$, then undergoing the phase transition in the strong coupling regime it becomes intrinsically non-perturbative QCD, which has a physical mass gap and confines gluons.*

Some important features of intrinsically non-perturbative QCD are collected:

(i) Its full gluon propagator (4.2.1) converges to the expression (4.3.5) after the renormalization of the mass gap is performed. This expression is effectively valid in the whole q^2-momentum plane. However, it is suppressed in comparison with the full gluon propagator for perturbative QCD in the $q^2 \to \infty$ limit.
(ii) It has a physical mass gap.
(iii) It confines dressed gluons in asymptotic states.
(iv) It has no free gluons at all in its formalism.
(v) It is ultraviolet finite and apparently infrared renormalizable.
(vi) Accumulating all the severe infrared singularities due to the self-interactions of massless gluon modes, with the help of the mass gap, it effectively becomes Abelian gauge theory beyond the gluon sector. It is easy to understand that all the skeleton loop diagrams with the number of intrinsically non-perturbative QCD gluon propagators (4.5.1) more than the number of independent loop variables will be suppressed. Only those will survive in the $\epsilon \to 0$ limit, where these numbers are equal.

Other interesting features of this theory may be established after the explicit including of the quark degrees of freedom into its formalism in the nearest future. However, first in the subsequent part we will show how the QCD vacuum structure is to be investigated within the Yang–Mills version of intrinsically non-perturbative QCD at zero- and non-zero temperatures.

Concluding, a few remarks are in order. QED has only the perturbative-type infrared singularity, $1/q^2$. This is in agreement with the cluster property of the Wightman functions [150], that's correlation functions of observables. In QCD the explicit presence of the regularized mass gap, which are necessarily accompanied by severe infrared singularities $(q^2)^{-2-k}$, $k = 0, 1, 2, 3, \ldots$, apparently, will violate this property. In turn, this validates the Strocchi theorem [151], which allows such a severely singular behavior of the full gluon propagator in QCD. However, this is not a problem, since QCD has no physical observables. In perturbative QCD, in which the gluon propagator is as much singular as $1/q^2$ only, the cluster property will not be violated. On the other hand, in intrinsically non-perturbative QCD with such a singular behavior of the relevant gluon propagator (4.3.5) the situation with the Wightman functions is not clear. It can be clarified only after the solution of the quark confinement problem, and a realistic calculation of the various physical observables within intrinsically non-perturbative QCD. At the fundamental quark–gluon level only these remarks make sense about the correlation between the structure of the corresponding gluon propagator and the properties of the Wightman functions. However, in Part II of this book, we show how to correctly calculate some physical parameters in the pure Yang–Mills version of the QCD.

PART II

Applications of the Mass Gap

Chapter 6

Vacuum Energy Density in the Quantum Yang–Mills Theory

6.1 Introduction

One of the main dynamical characteristics of the QCD ground state is the bag constant. Its name comes from the famous bag models [152, 153], but its present understanding (and thus modern definition) is not connected to hadron properties. It is defined as the difference between the perturbative and the non-perturbative vacuum energy densities (VEDs) [61, 135, 136, 154, 155]. So, we can symbolically put

$$B = VED^{PT} - VED ,$$

where VED is the non-perturbative but 'contaminated' by the perturbative contributions — this is a full VED like the full gluon propagator. At the same time, we can continue as follows:

$$\begin{aligned} B &= VED^{PT} - VED = VED^{PT} - \left[VED - VED^{PT} + VED^{PT}\right] \\ &= VED^{PT} - \left[VED^{INP} + VED^{PT}\right] = -VED^{INP} > 0 , \end{aligned}$$

since the vacuum energy density is always negative. The bag constant is nothing but the intrinsically non-perturbative vacuum energy density, apart from the sign, by definition, and thus is free of the perturbative contributions (contaminations). The symbolic subtraction presented here includes the subtraction at the fundamental gluon level, described in previous Chapters, and two others at the hadronic level, i.e., when the gluon degrees of freedom should be integrated out. In order to consider it also as a physical characteristic of the QCD ground state, the bag constant correctly calculated should satisfy some other necessary requirements such as colorlessness, finiteness, gauge-independence, no imaginary part (stable vacuum), etc.

The main purpose of this Chapter is to formulate a formalism which makes it possible to calculate correctly the quantum part of the bag constant, using the effective potential approach for composite operators [156–158]. In particular, this Chapter aims to show how the above-mentioned subtractions are to be analytically made. On account of the confining effective charge (4.4.1), the bag constant has been numerically evaluated, satisfying all the necessary requirements mentioned above. Using further the trace anomaly relation [159–162], we will also develop a general formalism which allows one to relate the bag constant to another important non-perturbative characteristic of the QCD ground state — the gluon condensate [135]. Here, we do not use the weak coupling solution for the corresponding β-function, but instead we will use the confining one (4.4.3). Finally we present here our numerical result for the bag constant, which is in a good agreement with other phenomenological estimates of the gluon condensate [135, 163].

6.2 The vacuum energy density

The quantum part of the vacuum energy density is determined by the effective potential approach for composite operators [156]. In the absence of external sources the effective potential is nothing but the vacuum energy density. It is given in the form of the skeleton loop expansion containing all the types of the QCD full propagators and vertices, see Fig. 6.2.1. So each vacuum skeleton loop itself is a sum of an infinite number of the corresponding vacuum loops. The number of the vacuum skeleton loops goes with the power of the Planck constant, $\hbar$.

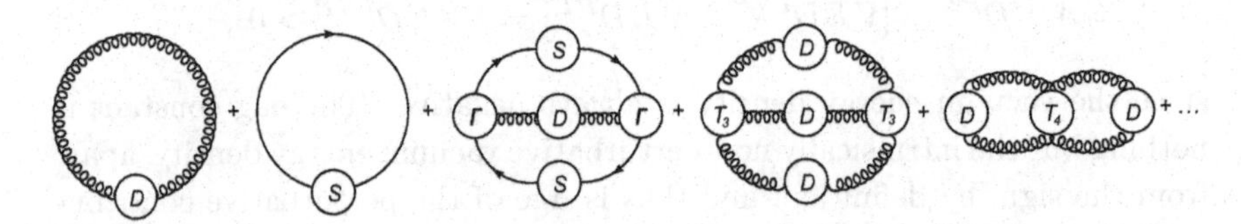

Fig. 6.2.1 The skeleton loop expansion for the effective potential. The wavy lines describe the full gluon propagators D. The solid lines describe the full quark propagators S. Γ is the full quark–gluon vertex, while T_3 and T_4 are the full three- and four-gluon vertices, respectively.

Here we are going to formulate a general method of numerical calculation of the quantum part of the intrinsically non-perturbative Yang – Mills

(YM) vacuum energy density in the covariant gauge QCD. The gluon part of the vacuum energy density to the leading order, the so-called log-loop level, $\sim \hbar$, is the first skeleton loop diagram in Fig. 6.2.1. This is analytically given by the effective potential for composite operators as follows [156]:

$$V(D) = \frac{i}{2} \int \frac{\mathrm{d}^4 q}{(2\pi)^4} \mathrm{Tr} \left\{ \ln \left(D_0^{-1} D \right) - \left(D_0^{-1} D \right) + 1 \right\}, \tag{6.2.1}$$

where $D(q)$ is the full gluon propagator and $D_0(q)$ is its free counterpart (see below). The traces over space-time and color group indices are assumed. Evidently, the effective potential is normalized to $V(D_0) = 0$, i.e., the free perturbative vacuum is normalized to zero, as usual. Next-to-leading and higher contributions (two and more vacuum skeleton loops) are suppressed at least by one order of magnitude in powers of $\hbar$. They generate very small numerical corrections to the log-loop terms, and thus are not important for the numerical calculation of the bag constant to the leading order. For the readers's convenience we will gather here all the necessary expressions, which are present in different Chapters of Part I, and remind that the signature is Euclidean.

The two-point Green's function, describing the full gluon propagator, is

$$D_{\mu\nu}(q) = i \left\{ T_{\mu\nu}(q) d(q^2; \xi) + \xi L_{\mu\nu}(q) \right\} \frac{1}{q^2}, \tag{6.2.2}$$

where $d(q^2; \xi)$ is the dimensionless gluon invariant function, the so-called Lorentz structure (sometimes, we will call it as the full gluon form factor or, equivalently, the running effective charge), while ξ is the gauge-fixing parameter and

$$T_{\mu\nu}(q) = \delta_{\mu\nu} - \frac{q_\mu q_\nu}{q^2} = \delta_{\mu\nu} - L_{\mu\nu}(q). \tag{6.2.3}$$

The free perturbative counterpart $D_0 \equiv D^0_{\mu\nu}(q)$ is obtained by putting the full gluon form factor $d(q^2; \xi)$ in equation (6.2.2) simply to one, i.e.,

$$D^0_{\mu\nu}(q) = i \left\{ T_{\mu\nu}(q) + \xi L_{\mu\nu}(q) \right\} \frac{1}{q^2}. \tag{6.2.4}$$

In order to evaluate the effective potential (6.2.1), on account of equations (6.2.2) and (6.2.4), we use the well-known expression

$$\mathrm{Tr} \ln \left(D_0^{-1} D \right) = 8 \cdot 4 \ln \det \left(D_0^{-1} D \right) = 32 \ln \left[\left(\frac{3}{4} \right) d\left(q^2; \xi \right) + \left(\frac{1}{4} \right) \right], \tag{6.2.5}$$

which becomes zero indeed when setting $d(q^2; \xi) = 1$.

Substituting these expressions into equation (6.2.1) and doing some algebra, one obtains ($\epsilon_g = V(D)$)

$$\epsilon_g = -16 \int \frac{d^4 q}{(2\pi)^4} \left[\ln \left[1 + 3d(q^2; \xi) \right] - \frac{3}{4} d\left(q^2; \xi\right) + a \right], \tag{6.2.6}$$

where constant $a = (3/4) - 2\ln 2 = -0.6363$ and the integration from zero to infinity over q^2 is assumed. The vacuum energy density ϵ_g derived in equation (6.2.6) is already a colorless quantity, since it has been summed over color indices. Also it does not depend explicitly on the unphysical (longitudinal) part of the full gluon propagator due to the product $(D_0^{-1} D)$, which, in its turn, comes from the above-mentioned normalization to zero. Thus it is worth emphasizing that the transversal (physical) degrees of freedom of the gauge bosons contribute only to this equation. Note, in the effective potential approach there is no need for ghost degrees of freedom in order to cancel the longitudinal (unphysical) component of the full gluon propagator. This originates from the normalization condition, which mathematically normalizes the free perturbative vacuum to zero and, in parallel, has physical meaning as well. Let us note that an overall numerical factor $1/2$ has been introduced into equation (6.2.1) in order to make the gluon degrees of freedom equal $32/2 = 16 = 8 \times 2$, where 8 colors of gluons times 2 helicity (transversal) degrees of freedom, see equations (6.2.5) and (6.2.6). That is why the ghost skeleton loops are not shown in Fig. 6.2.1.

In the connection with the above-mentioned normalization condition a few remarks are in order. It does not work for the higher order vacuum loops. As explained in Ref. [156], for consistency with them in the perturbative QCD Green's functions, for example in the Hartree–Fock approximation, the Landau gauge should be used. In the Ref. [164] the effective potential has been used to the two-loop order for the investigation of QCD chiral-symmetry breaking just in the Landau gauge and in the Hartree–Fock approximation. In the general case (beyond perturbation theory and at any gauge), however, the cancelation of unphysical gluon modes should proceed with the help of ghosts as it is described in more detail in Appendix 6.A.

The derived expression (6.2.6) remains rather formal, since it suffers from two serious problems: the coefficient of the transversal Lorentz structure $d(q^2; \xi)$ may still depend explicitly on ξ. Furthermore, it is divergent at least as the fourth power of the ultraviolet cutoff, and therefore suffers from different types of the perturbative contributions.

6.3 The intrinsically non-perturbative vacuum energy density

In order to define the vacuum energy density free of the above-mentioned perturbative contributions ('contaminations'), let us make first the subtraction at the fundamental gluon level, namely

$$d\left(q^2;\xi\right) = d\left(q^2;\xi\right) - d^{PT}\left(q^2;\xi\right) + d^{PT}\left(q^2;\xi\right) = d^{INP}\left(q^2\right) + d^{PT}\left(q^2;\xi\right), \tag{6.3.1}$$

in complete agreement with the subtraction at the gluon propagator level in Section (2.3.4). Here $d^{PT}(q^2;\xi)$ correctly describes the perturbative structure of the full effective charge $d(q^2;\xi)$, including its behavior in the ultraviolet limit (asymptotic freedom), otherwise remaining arbitrary. On the other hand, $d^{INP}(q^2)$ defined by the above-made subtraction, is assumed to reproduce correctly the intrinsically non-perturbative structure of the full effective charge, including its asymptotic in the deep infrared limit. This underlines the strong intrinsic influence of the infrared properties of the theory on its intrinsically non-perturbative dynamics. Evidently, both terms are valid in the whole energy/momentum range, i.e, they are not asymptotics, and they are uniquely and exactly separated from each other with respect of the mass gap Δ_R^2, as has been already described in the previous Chapters. In principle, in some special models of the QCD vacuum, such as the Abelian Higgs model [165], the non-perturbative scale is to be identified with the mass of the dual gauge boson.

Substituting the decomposition (6.3.1) into equation (6.2.6) and doing some simple rearrangements, one obtains

$$\epsilon_g = -\frac{1}{\pi^2}\int \mathrm{d}q^2\, q^2 \left[\ln\left[1 + 3d^{INP}\left(q^2\right)\right] - \frac{3}{4}d^{INP}\left(q^2\right)\right] + \epsilon_{PT}, \tag{6.3.2}$$

where the trivial integration over the angular variables in equation (6.2.6) has been already done. Here ϵ_{PT} is

$$\epsilon_{PT} = -\frac{1}{\pi^2}\int \mathrm{d}q^2\, q^2 \left[\ln\left[1 + \frac{3d^{PT}\left(q^2;\xi\right)}{1 + 3d^{INP}\left(q^2\right)}\right] - \frac{3}{4}d^{PT}\left(q^2;\xi\right) + a\right]. \tag{6.3.3}$$

It contains the contribution which is mainly determined by the perturbative part of the full effective charge, $d^{PT}\left(q^2,\xi\right)$. The constant a should be also included, since it comes from the normalization of the free perturbative vacuum to zero.

However, this is not the whole story yet. The first term in equation (6.3.2), depending only on the intrinsically non-perturbative effective charge, nevertheless, assumes the integration over the perturbative region (up to infinity). It also represents the type of the perturbative contribution, which should be subtracted as well. If we separate the non-perturbative region from the perturbative one, by introducing the so-called effective scale q^2_{eff} explicitly, then we get

$$\epsilon_g = -\frac{1}{\pi^2} \int_0^{q^2_{eff}} \mathrm{d}q^2\, q^2 \left[\ln\left[1 + 3d^{INP}\left(q^2\right)\right] - \frac{3}{4} d^{INP}\left(q^2\right) \right] + \epsilon_{PT} + \epsilon'_{PT}\,, \tag{6.3.4}$$

where

$$\epsilon'_{PT} = -\frac{1}{\pi^2} \int_{q^2_{eff}}^{\infty} \mathrm{d}q^2\, q^2 \left[\ln\left[1 + 3d^{INP}\left(q^2\right)\right] - \frac{3}{4} d^{INP}\left(q^2\right) \right]. \tag{6.3.5}$$

This integral represents the contribution to the vacuum energy density which is determined by the intrinsically non-perturbative part of the full gluon propagator but integrated out over the perturbative region. Along with ϵ_{PT} given in equation (6.3.3) it also represents a type of the perturbative contribution into the gluon part of the vacuum energy density (6.3.2), as mentioned above. This means that the two remaining terms in equation (6.3.4) should be subtracted by introducing the intrinsically non-perturbative Yang–Mills vacuum energy density ϵ_{YM} as follows:

$$\epsilon_{YM} = \epsilon_g - \epsilon_{PT} - \epsilon'_{PT}, \tag{6.3.6}$$

where the explicit expression for ϵ_{YM} is given by the integral in equation (6.3.4).

Concluding, let us emphasize that both subtracted terms ϵ_{PT} and ϵ'_{PT}, strictly speaking, are not purely perturbative, since along with the non-trivial perturbative effective charge $d^{PT}(q^2)$ they contain the intrinsically non-perturbative effective charge $d^{INP}(q^2)$ as well. So to call them the perturbative contributions is a convention. More precisely it is better to say that these terms are contaminated by the perturbative contributions. The above-mentioned necessary subtractions can be made in a more sophisticated way by introducing explicitly the ghost degrees of freedom (see Appendix 6.A.

6.4 The bag constant

The bag constant (or so-called bag pressure) is defined as the difference between the perturbative and the non-perturbative vacuum energy densities [61, 135, 136, 154, 155]. So in our notations for the Yang – Mills fields, and as it follows from the definition by equation (6.3.6), it is nothing but the intrinsically non-perturbative Yang – Mills vacuum energy density apart from the sign, i.e.,

$$\begin{aligned} B_{YM} &= -\epsilon_{YM} = \epsilon_{PT} + \epsilon'_{PT} - \epsilon_g \\ &= \frac{1}{\pi^2} \int_0^{q^2_{eff}} \mathrm{d}q^2 \, q^2 \left[\ln\left[1 + 3\alpha^{INP}\left(q^2\right)\right] - \frac{3}{4}\alpha^{INP}\left(q^2\right) \right], \end{aligned} \tag{6.4.1}$$

where from now on we introduce the notation $d^{INP}\left(q^2\right) \equiv \alpha^{INP}\left(q^2\right)$. This is a general expression for any model effective charge in order to calculate the bag constant, or the intrinsically non-perturbative Yang – Mills vacuum energy density apart from the sign, from first principles. It is our definition of it and thus of the bag constant. So it is defined as the special function of the intrinsically non-perturbative effective charge integrated out over the non-perturbative, soft momentum region, where $0 \leq q^2 \leq q^2_{eff}$. It is free of the perturbative contributions by construction. In this connection, let us recall that ϵ_g is also non-perturbative, but contaminated by the perturbative contributions (see equation (6.3.4), which need just to be subtracted in order to get equation (6.4.1). Comparing expressions (6.2.6) and (6.4.1), one comes to the following 'prescription' to get equation (6.4.1) directly from equation (6.2.6):

(i) Replacing $d\left(q^2\right) \rightarrow d^{INP}\left(q^2\right)$ or, equivalently, $\alpha\left(q^2\right) \rightarrow \alpha^{INP}\left(q^2\right)$.
(ii) Omitting the constant a which normalizes the free perturbative vacuum to zero.
(iii) Introducing the effective scale q^2_{eff} which separates the non-perturbative region from the perturbative one in the q^2-momentum space.
(iv) Omitting the minus sign for the bag constant.

At this stage the bag constant defined by equation (6.4.1) is definitely colorless (color-singlet) and free of perturbative contaminations. Let us remind that it also depends on only transversal degrees of freedom of gauge bosons (gluons). All its other properties mentioned above (finiteness, positivity, no imaginary part, etc.) depend on the chosen effective charge, more precisely on its intrinsically non-perturbative counterpart. It is worth emphasizing once more that in defining correctly the intrinsically non-perturbative Yang–Mills vacuum energy density — or, equivalently, the bag constant — three types of the corresponding subtractions have been introduced. The first one is in equation (6.3.1) at the fundamental gluon level and the two others are in equation (6.3.6), when the gluon degrees of freedom were to be integrated out.

For actual numerical calculations of the bag constant via the expression (6.4.1) we are going to use the intrinsically non-perturbative effective charge (4.4.1), and introducing the dimensionless quantities, namely

$$\alpha^{INP}\left(q^2\right) = \alpha^{INP}\left(z\right) = \frac{z_c}{z}, \text{ where } z = \frac{q^2}{q^2_{eff}} \text{ and } z_c = \frac{\Delta^2_R}{q^2_{eff}}. \quad (6.4.2)$$

On the other hand it is always convenient to factorize its scale dependence as follows:

$$B_{YM}(q^2_{eff}) = q^4_{eff} \cdot \Omega_{YM}, \quad (6.4.3)$$

where we introduce the dimensionless intrinsically non-perturbative Yang–Mills effective potential Ω_{YM}, for convenience. Its explicit expression is

$$\Omega_{YM} = \frac{1}{\pi^2} \int_0^1 \mathrm{d}z\ z \left[\ln\left[1 + 3\alpha^{INP}(z)\right] - \frac{3}{4}\alpha^{INP}(z)\right]. \quad (6.4.4)$$

Let us emphasize that in order to factorize the scale dependence in the effective potential it is necessary to choose the fixed scale, like q^2_{eff}, and not the scale which can be varied, for example like the mass gap which can go to zero in order to recover the perturbation theory limit. Equations (6.4.2), (6.4.3), and (6.4.4) are the main subject of our consideration in what follows. It is worth emphasizing once more that these expressions are general ones in order to correctly calculate the bag constant from first principles, thus to get the color-singlet expression, which is free of all the perturbative contributions.

6.5 Analytical and numerical evaluation of the bag constant

In terms of the dimensionless quantities of equation (6.4.2) the dimensionless effective potential (6.4.4) becomes

$$\Omega_{YM}(z_c) = \frac{1}{\pi^2}\int_0^1 dz\ z\left\{\ln\left[1+\left(\frac{3z_c}{z}\right)\right] - \frac{3}{4}\frac{z_c}{z}\right\}. \qquad (6.5.1)$$

Performing an almost trivial integration in this integral, one obtains

$$\Omega_{YM}(z_c) = \frac{1}{2\pi^2}z_c^2\left[\frac{3}{2z_c} + \frac{1}{z_c^2}\ln(1+3z_c) - 9\ln\left(1+\frac{1}{3z_c}\right)\right]. \qquad (6.5.2)$$

It is easy to see now that the effective potential (6.5.2) approaches zero at the $z_c \to 0$ limit. At infinity $z_c \to \infty$ it diverges at around $-z_c$. From equation 6.4.2 it follows at a fixed effective scale, q^2_{eff} that, $z_c \to 0$ is a correct perturbative regime, while $z_c \to \infty$ is not a physical regime, since the mass gap Δ^2 is either finite or zero (the perturbation theory limit), thus it cannot be infinitely large. In other words, at a fixed effective scale one recovers the correct perturbative limit for the bag constant — the above-mentioned normalization condition is maintained for the bag constant, as it should be.

The nontrivial second zero of the effective potential (6.5.2) follows obviously from the condition,

$$3z_c + 2\ln(1+3z_c) - 18z_c^2\ln\left[1+\left(\frac{1}{3z_c}\right)\right] = 0, \qquad (6.5.3)$$

the numerical solution of which is

$$z_c^0 = 1.3786. \qquad (6.5.4)$$

Evidently, through the relation (6.4.2) this value determines a possible upper bound for Δ_r^2 and lower bound for q^2_{eff}, since B_{YM}/ϵ_{YM} is always positive/negative, see Figs. 6.5.1 and 6.5.2.

At $z_c = 0$, $\Delta_R^2 = 0$, thus the effective potential (6.5.2) vanishes identically, as it should be. From the above one can conclude that this effective potential as a function of z_c has a maximum at some finite point, see Fig. 6.5.1. In the way it has been introduced z_c plays the role of the constant of integration of the effective potential though being formally a parameter of the theory. In general, by taking the first derivative of the effective potential with respect to the constant of integration one recovers the corresponding

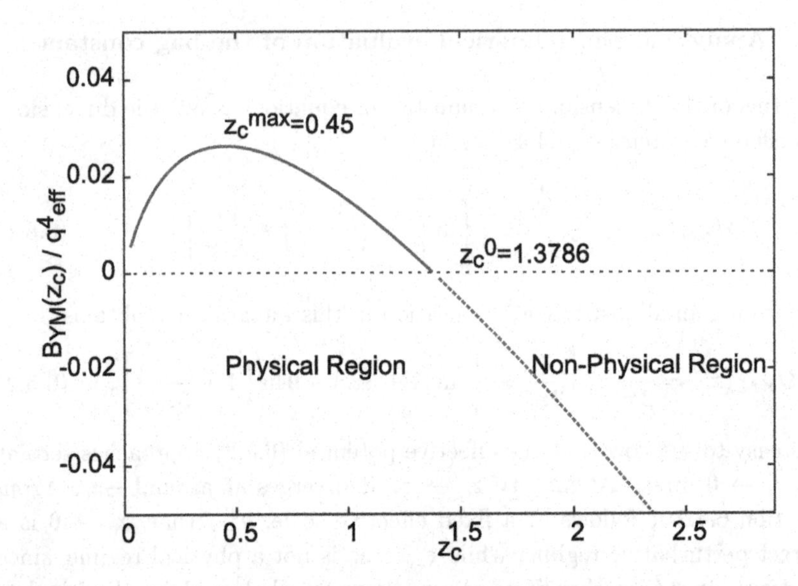

Fig. 6.5.1 The B_{YM}/q^4_{eff} effective potential vs. z_c. The non-physical region is $z_c \geq z^0_c$, since B_{YM} should be always positive. At $z_c = 0$ the effective potential is also zero.

equations of motion [156–158]. Requiring thus $\partial\Omega_{YM}(z_c)/\partial z_c = 0$, one obtains:

$$z_c^{-1} = 4 \ln \left[1 + \left(\frac{1}{3z_c}\right)\right] , \tag{6.5.5}$$

which makes it possible to fix the constant of integration of the corresponding equation of motion at maximum. Its numerical solution is

$$z_c^{max} = 0.4564, \tag{6.5.6}$$

so at maximum the ratio Δ^2_R/q^2_{eff} is always less than one. At this point the numerical value of the effective potential (6.5.2) is

$$\Omega_{YM}(z_c^{max}) = \frac{1}{2\pi^2}\left[\frac{3}{4}z_c^{max} - \ln\left(1 + 3z_c^{max}\right)\right] = 0.0263. \tag{6.5.7}$$

The bag constant defined in equation (6.4.3), and hence the corresponding intrinsically non-perturbative vacuum energy density (6.4.1), as a function of q^4_{eff} or, equivalently, of the mass gap Δ^4_R thus becomes,

$$B_{YM} = -\epsilon_{YM} = 0.0263\, q^4_{eff} = 0.1273\ \Delta^4_R, \tag{6.5.8}$$

where the relation

$$q^2_{eff} = (z_c^{max})^{-1}\Delta^2_R = 2.2\ \Delta^2_R \tag{6.5.9}$$

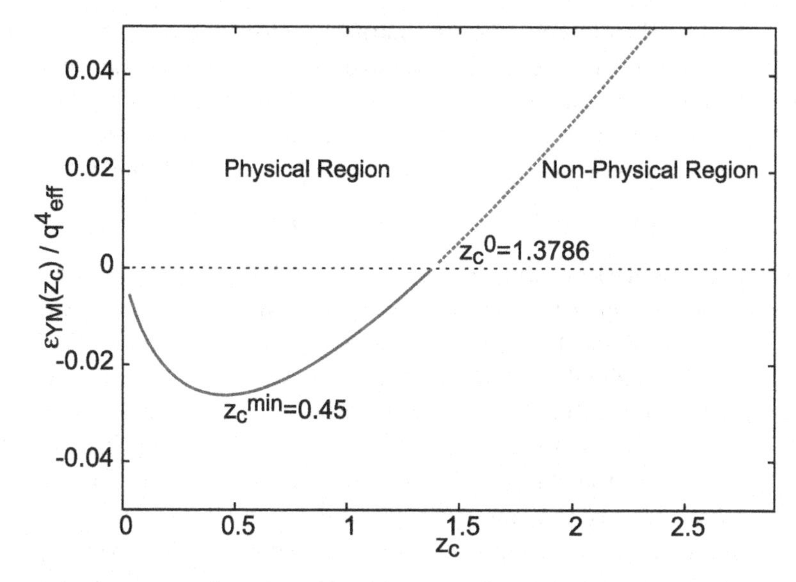

Fig. 6.5.2 The ϵ_{YM}/q^4_{eff} effective potential vs. z_c. The non-physical region is $z_c \geq z^0_c$, since ϵ_{YM} should be always negative. At $z_c = 0$ the effective potential is also zero.

has been already used. It is worth noting that a maximum for the bag constant corresponds to a minimum for the intrinsically non-perturbative Yang – Mills vacuum energy density ϵ_{YM} — the so-called 'stationary' state, see Fig. 6.5.2.

So, here we have explicitly demonstrated that in the considered case the bag constant (6.5.8) is finite, positive, and it has no imaginary part, indeed. It depends only on the mass gap responsible for the intrinsically non-perturbative dynamics in the QCD ground state or, equivalently, on the effective scale squared separating the non-perturbative region from the perturbative one.

Scale-setting schemes and numerical results

In order to complete the numerical calculation of the above defined bag constant all we need now is the value for the effective scale q^2_{eff}, which separates the non-perturbative region from the perturbative one. Similarly, the value for a scale at which the non-perturbative effects become important, that is the mass gap Δ^2_R, also allows one to achieve the same goal. If the perturbative regime for gluons (as well as for quarks) starts conventionally from ~ 1 GeV, then this number is a natural choice for the effective scale. It makes it also possible to directly compare our values with the values of many phenomenological parameters calculated just at this scale. We consider this value as a well justified and realistic upper limit for the effective scale defined above. Thus, using further the relation (6.5.9), one gets

$$q_{eff}^2 = 1 \text{ GeV}^2 \quad \text{and} \quad \Delta_R^2 = 0.4564 \text{ GeV}^2. \tag{6.5.10}$$

Similarly, the numerical value of the mass gap Δ_R^2 has been obtained from the experimental value for the pion decay constant, $F_\pi = 93.3$ MeV, by implementing a physically well-motivated scale-setting scheme [102, 166]. In fact, we approximate the pion decay constant in the chiral limit F_π^0 by its experimental value, since the difference between them can be a few MeV only. This is due to the smallness of the corresponding light quark current masses. In the chiral limit the difference between the Yang–Mills version for the scale of the non-perturbative dynamics, that is Δ_R^2, and the corresponding QCD scale, that is Λ_{INP}^2 introduced in the previous Chapter, is also expected to be small enough. The pion decay constant is a good experimental number, since it is a directly measured quantity, contrary to for example the quark condensate or the dynamically generated quark mass. For the mass gap we have obtained the following numerical result: $\Delta_R = 0.5784$ GeV, so similarly to the relation (6.5.10), one yields

$$\Delta_R^2 = 0.3345 \text{ GeV}^2 \quad \text{and} \quad q_{eff}^2 = 0.733 \text{ GeV}^2. \tag{6.5.11}$$

In what follows we will consider this value as a realistic lower limit for the effective scale. One has to conclude that we have obtained rather close numerical results for the effective scale and the mass gap, by implementing rather different scale-setting schemes. It is worth emphasizing that the effective scale (6.5.11) covers quite well not only the deep infrared region but the substantial part of the intermediate one as well.

For the above-mentioned possible upper bounds for Δ_R^2 and lower bounds for q_{eff}^2 our numerical results are for the scale-setting scheme (6.5.10):

$$\Delta_R^2 \leq 1.379 \text{ GeV}^2 \quad \text{and} \quad q_{eff}^2 \geq 0.330 \text{ GeV}^2 \tag{6.5.12}$$

then similarly, based on the scale-setting scheme (6.5.11):

$$\Delta_R^2 \leq 1.01 \text{ GeV}^2 \quad \text{and} \quad q_{eff}^2 \geq 0.242 \text{ GeV}^2 \tag{6.5.13}$$

Evidently, their calculated values in each scale-setting scheme satisfy the corresponding bounds.

For the bag constant — and hence for the intrinsically non-perturbative Yang – Mills vacuum energy density — from equation (6.5.8), one obtains

$$B_{YM} = -\epsilon_{YM} = (0.0142 - 0.0263)\ \mathrm{GeV}^4, \tag{6.5.14}$$

where the first and second numbers in brackets correspond to the numerical values given in equations (6.5.10) and (6.5.11), respectively.

Concluding, let us note that in the pure Yang – Mills theory there is no way to calculate the mass gap independently of the well-motivated scale-setting scheme, that's the effective scale in this case, i.e., relations (6.5.10). The scale-setting scheme (6.5.11) is based on the numerical value of the pion decay constant in the chiral limit. So this scheme is legitimate to use here as well, since in the chiral limit the quark degrees of freedom do not contribute to the vacuum energy density, as it follows from the trace anomaly relation (see the next Section).

6.6 The trace anomaly relation

The intrinsically non-perturbative vacuum energy density — and hence the bag constant — is important by itself as the main dynamical characteristic of the QCD ground state. Furthermore it assists in calculating such an important phenomenological parameter as the gluon condensate, introduced in the QCD sum rules approach to the physics of resonances [135]. The famous trace anomaly relation [159–162] in the general case of non-zero current quark masses m_f^0 is

$$\Theta_{\mu\mu} = \frac{\beta(\alpha_s)}{4\alpha_s}\, G^a_{\mu\nu}\, G^a_{\mu\nu} + \sum_f m_f^0 \overline{q}_f q_f, \tag{6.6.1}$$

where $\Theta_{\mu\mu}$ is the trace of the energy–momentum tensor and $G^a_{\mu\nu}$ is the gluon field strength tensor, while for the ratio $\beta(\alpha_s)/\alpha_s$ see discussion below[1]. The trace anomaly relation which includes the anomalous dimension for the quark mass has been derived in [162], however, in our case of the pure gluon fields we can use the standard form of the trace anomaly relation (6.6.1). Sandwiching it between vacuum states and taking into account the obvious relation $\langle 0\,|\Theta_{\mu\mu}|\,0\rangle = 4\epsilon_t$, one obtains

$$4\epsilon_t = \langle 0|\frac{\beta\,(\alpha_s)}{4\alpha_s}\, G^a_{\mu\nu}\, G^a_{\mu\nu}|0\rangle + \sum_f m_f^0\ \langle 0|\overline{q}_f q_f|0\rangle. \tag{6.6.2}$$

[1]Let us note that here and in the next Section we have temporarily restored the subscript 's' in the expressions for the effective charge and its corresponding β-function in order to comply with the standard expression for the trace anomaly.

Here ϵ_t is the sum of all possible independent non-perturbative contributions to the vacuum energy density (the total vacuum energy density) and $\langle 0 | \overline{q}_f q_f | 0 \rangle$ is the chiral quark condensate. From this equation in the case of the pure Yang–Mills fields (i.e., when the number of quark fields is zero, $N_f = 0$), one can get

$$\langle 0 | \frac{\beta(\alpha_s)}{4\alpha_s} G^a_{\mu\nu} G^a_{\mu\nu} | 0 \rangle = 4\, \epsilon_{YM}, \tag{6.6.3}$$

where the total vacuum energy density ϵ_t can be saturated by the intrinsically non-perturbative Yang–Mills vacuum energy density, ϵ_{YM}, defined in equation (6.4.1). Thus one can set $\epsilon_t = \epsilon_{YM} = -\epsilon_{PT} - \epsilon'_{PT} + \epsilon_g$. Let us note that the same result by equation (6.6.3), will be obtained in the chiral limit for light quarks $m^0_f = 0$, for $f = 1, 2, 3$ as well.

If confinement happens then the β-function is always in the domain of attraction (which is always negative) without any infrared stable fixed points [2]. Therefore, it is convenient to introduce the general definition of the gluon condensate not using the weak coupling limit solution to the β-function as follows:

$$\langle G^2 \rangle \equiv -\langle 0 | \frac{\beta(\alpha_s)}{4\alpha_s} G^a_{\mu\nu} G^a_{\mu\nu} | 0 \rangle = -4\, \epsilon_{YM} = 4\, B_{YM}. \tag{6.6.4}$$

Thus, the above defined general gluon condensate will be always positive, as it should be. The importance of this relation is that it gives the value of the gluon condensate as a function of the bag constant whatever the solution of the β-function in terms of α_s is. However, let us remind that there is a correlation between the two sides of this equation. The bag constant, correctly defined in equation (6.4.1), depends, in general, on the intrinsically non-perturbative effective charge $\alpha_s^{INP}(q^2)$. On the other hand, the renormalization group equation

$$q^2 \frac{d\alpha_s(q^2)}{dq^2} = \beta\left(\alpha_s(q^2)\right) \tag{6.6.5}$$

for the β-function gives it in terms of the corresponding effective charge. This makes it possible to determine the ratio, $\beta(\alpha_s)/\alpha_s \equiv \beta\left(\alpha_s(q^2)\right)/\alpha_s(q^2)$, which appears in the left-hand-side of equation (6.6.4). Of course, this equation should be solved for the chosen intrinsically non-perturbative effective charge (see the next Section).

Concluding, let us only note that the quantum part of the total intrinsically non-perturbative vacuum energy density at log-loop level is:

$$\epsilon_t = \epsilon_{YM} + N_f\, \epsilon_q, \tag{6.6.6}$$

where ϵ_q is the intrinsically non-perturbative quark skeleton loop contribution for its skeleton loop diagram as it is shown on Fig. 6.2.1. It is an order of magnitude less than ϵ_{YM} because of much less quark degrees of freedom in the vacuum, and it is positive because of overall minus due to the quark loop. Evidently, in terms of the Yang – Mills bag constant, one obtains

$$\epsilon_t = -B_{YM}\left[1 - \nu N_f\right], \tag{6.6.7}$$

where we introduce $\epsilon_q = \nu B_{YM}$ and $\nu \ll 1$. So the replacement of the total bag constant by its Yang – Mills counterpart only is a rather good approximation from the numerical point of view. In this connection, let us remind that in the large N_c-limit the pure gluon contribution scales as N_c^2, while the quark contribution scales only as N_c [71]. In order to correctly calculate the bag constant in full QCD the quark part of the intrinsically non-perturbative vacuum energy density is also important, but is beyond the scope of the pure Yang – Mills applications of the mass gap present in Part II of this book.

6.7 Comparison with phenomenology

Let us show explicitly now that our numerical values for the bag constant calculated in (6.5.14)) are in a rather good agreement with the phenomenological values of the gluon condensate. Above we have already developed a general formalism which allows one to express the gluon condensate as a function of the bag constant. So substituting the numerical value of the bag constant into the equation (6.6.4), one obtains:

$$\langle G^2 \rangle \equiv -\langle 0 | \frac{\beta(\alpha_s)}{4\alpha_s} G^a_{\mu\nu} G^a_{\mu\nu} | 0 \rangle = 4\, B_{YM} = (0.0568 - 0.1032) \ \mathrm{GeV}^4. \tag{6.7.1}$$

On the other hand, the renormalization group equation for the β-function given by (6.6.5) after the substitution of our solution for the intrinsically non-perturbative effective charge (4.4.1) yields:

$$\beta(\alpha_s(q^2)) = -\alpha_s(q^2), \tag{6.7.2}$$

in complete agreement with equation (4.4.3). The corresponding ratio as it appears in the left-hand-side of equation (6.7.1) is

$$\frac{\beta(\alpha_s)}{\alpha_s} \equiv \frac{\beta(\alpha_s(q^2))}{\alpha_s(q^2)} = -1, \tag{6.7.3}$$

again in complete agreement with equation (4.4.4) as it is expected. Substituting further this solution into the equation (6.7.1), it becomes

$$\langle 0|\frac{1}{4}G^a_{\mu\nu}G^a_{\mu\nu}|0\rangle = 4\,B_{YM} = (0.0568 - 0.1052)\ \mathrm{GeV}^4, \tag{6.7.4}$$

which means that both sides of this relation between the bag constant and the gluon condensate have been calculated by using the same expression for the intrinsically non-perturbative effective charge, and hence for the corresponding β-function. So from the numerical point of view the bag constant and the gluon condensate are in a self-consistent dependence from each other, making thus the latter free of all types of perturbative contributions. Our expression for the gluon condensate (6.7.4) allows one to recalculate any gluon condensate at any scale and any ratio, $\beta(\alpha_s)/\alpha_s$. To the gluon condensate (6.41) a physical meaning can indeed be assigned as the global (average) vacuum characteristic which measures a density of the purely transversal infrared severely singular (i.e., intrinsically non-perturbative) virtual gluon fields configurations in the QCD vacuum.

However, it cannot be directly compared with the phenomenological values for the standard gluon condensate estimated within different approaches [163]. The problem is that it is necessary to remember that any value at the scale as in equation (6.5.11) (lower bound in the right-hand-side of equation (6.7.4)) is to be recalculated at the 1 GeV scale. Moreover, both values explicitly shown in equation (6.7.5) should be recalculated at the same ratio, as mentioned above.

In phenomenology the standard ratio of the gluon condensate and its numerical value is:

$$G_2 = \langle \frac{\alpha_s}{\pi} G^2 \rangle = \langle 0| \frac{\alpha_s}{\pi} G^a_{\mu\nu}G^a_{\mu\nu}|0\rangle \approx 0.012\ \mathrm{GeV}^4\ , \tag{6.7.5}$$

which can be changed within a factor of ~ 2 [135] (let us recall that this ratio comes from the weak coupling solution for the β function, see for example in [149]). Thus in order to achieve the same parameterization both sides of equation (6.7.4) should be multiplied by $4(\alpha_s/\pi)$. For the numerical value of the strong fine structure constant we use $\alpha_s = \alpha_s(m_Z) = 0.1187$ from the Particle Data Group [167]. In addition, the lower bound should be multiplied by the factor $(1/0.733)^2 = 1.86$, coming from the numerical value by equation (6.5.11). Then the recalculated gluon condensate in (6.7.4), which is denoted as $\bar{G}_2$, finally becomes

$$\bar{G}_2 = 0.016\ \mathrm{GeV}^4, \tag{6.7.6}$$

i.e., both numerical numbers in equation (6.7.4) coincide as it should be. This numerical value for the gluon condensate should be compared with the numerical value coming from the phenomenology, see equation (6.7.5) above. This shows that all our numerical results are in good agreement with various phenomenological estimates [135, 163]. This confirms that our numerical values for the bag constant, and hence for the gluon condensate, are rather realistic ones.

6.8 Numerical values for B_{YM} in different units

In order to show explicitly what magnitude of numbers we are dealing with, let us present our numerical value for the bag constant given by equation (6.5.14) in different units, namely:

$$\begin{aligned} B_{YM} = -\epsilon_{YM} &= (0.0142 - 0.0263)\ \mathrm{GeV}^4 \\ &= (1.84 - 3.4)\ \mathrm{GeV/fm}^3 \\ &= (1.84 - 3.4) \cdot 10^{39}\ \mathrm{GeV/cm}^3. \end{aligned} \tag{6.8.1}$$

This is a huge amount of energy stored in one cm^3 of the QCD vacuum even in 'God-given' units $\hbar = c = 1$. Using the number of different conversion factors (see, for example [3] or the Particle Data Group [167]) the bag constant can be expressed in different systems of units (SI, CGS, etc.).

Taking further into account that

$$1\ \mathrm{GeV} = 1.6 \times 10^{-10}\mathrm{J} = 4.45 \times 10^{-23}\ \mathrm{GWh}, \tag{6.8.2}$$

from equation (6.8.1) one gets ($1\ \mathrm{W} = 10^{-3}\ \mathrm{kW} = 10^{-6}\ \mathrm{MW} = 10^{-9}\ \mathrm{GW}$)

$$B_{YM} = (8.2 - 15) \cdot 10^{16}\ \mathrm{GWh/cm}^3 \tag{6.8.3}$$

or, equivalently,

$$E_{YM} = B_{YM}\ \mathrm{cm}^3 = (8.2 - 15) \cdot 10^{16} \sim 10^{17}\ \mathrm{GWh} \tag{6.8.4}$$

in familiar units of watt-hour (Wh). Let us note that if one puts the effective scale squared as small as realistically possible $q^2_{eff} = 0.242\ \mathrm{GeV}^2$ (see equation (6.5.13)), then the previous number will be only slightly changed, namely $E_{YM} = B_{YM}\ \mathrm{cm}^3 = (4.8 - 8.7) \times 10^{15}$ GWh. So both numbers still indicate a huge amount of the bag energy E_{YM} stored in one cm^3 of the QCD vacuum.

It is especially interesting to compare these numbers with the total production of primary energy of the 25 European Union member countries in year 2004 which was [168] (see also [169])

$$E_t \sim 10.2\text{PWh} = 10.2 \cdot 10^6 \text{ PWh} \sim 10^7 \text{ GWh}. \tag{6.8.5}$$

where 1 PWh = 1 Petawatt-hour. Approximately 1/3 of this energy was produced by nuclear power plants [168, 169]. The huge difference between the numbers in equations (6.8.4) and (6.8.5) is very impressive and leads to some interesting, still speculative, but already possible discussion below and in our preliminary work [170].

6.9 Contribution of B_{YM} to the dark energy problem

Apparently, our bag constant (6.8.1) may also contribute to the so-called dark energy density [171]. At least, from the qualitative point of view it satisfies almost all the criteria necessary for the dark energy/matter candidate as seen in Section 6.10 and the discussions in [171, 172]. From the quantitative numerical point of view it is also much better than the estimate from the Higgs field's contribution to the vacuum energy density, which is about [173, 174]

$$\varrho_H \sim 10^8 \text{ GeV}^4. \tag{6.9.1}$$

In this notation our value (6.8.1) is about

$$\varrho_{our} \approx 10^{-2} \text{ GeV}^4. \tag{6.9.2}$$

The observed vacuum energy density is very small indeed, namely

$$\varrho_{vac} \approx 10^{-46} \text{ GeV}^4, \tag{6.9.3}$$

see [173–175]. So relatively to the value inferred from the cosmological constant – the above-mentioned observed vacuum energy density.

$$\varrho_H / \varrho_{vac} \sim 10^{54}, \tag{6.9.4}$$

while ours is

$$\varrho_{our} / \varrho_{vac} \sim 10^{44}, \tag{6.9.5}$$

i.e, some ten orders of magnitude better than is expected from the direct comparison of the estimate (6.9.1) with our value (6.9.2).

Let us note that calculating at the Planck length scale [167], we will obtain the same ratio, as it should be. From equation (6.8.1) it follows that

$$\varrho_{our} \sim 10^{39} \text{ GeV/cm}^3 = 10^{-60} \text{ GeV}/L_p^3, \tag{6.9.6}$$

where we used cm $= 10^{33}\ L_p$ and L_p denotes the above-mentioned Plank length [167]. In these units the observed vacuum energy density is

$$\varrho_{vac} \sim 10^{-46}\ \mathrm{GeV}^4 \sim 10^{-5}\ \mathrm{GeV/cm}^3 = 10^{-104}\ \mathrm{GeV}/L_p^3, \qquad (6.9.7)$$

so that the ratio between (6.9.6) and (6.9.7) becomes again (6.9.5), indeed. Of course, the ratio (6.9.5) still remains very large, but it is much better than the ratio (6.9.4), as emphasized above. Other possibilities of how QCD can be related to the dark energy puzzle has been described in [176] and references therein.

Concluding, the vacuum for which the value (6.9.3) has been measured should not be mixed with the vacuum of any quantum field gauge theory. For the former its energy is always positive, so the vacuum is simply treated as an empty space. The energy of the latter is always negative, and it is full of any kind of quantum excitations, fluctuations, etc. However, the QCD bag constant is always positive, finite, gauge-invariant, etc. — if it has been correctly defined and calculated like here. That is the primary reason why we can compare our value (6.9.2) and the estimates (6.9.1) with (6.9.3).

6.10 Energy from the QCD vacuum

The Lamb shift and the Casimir effect are probably the two most famous experimental evidence of zero-point energy fluctuations in the vacuum of QED [92, 177–179]. Both effects are rather weak, since the QED vacuum is mainly perturbative by origin, character and magnitude (the corresponding fine structure constant is weak). However, even in this case attempts have been already made to exploit the Casimir effect in order to 'observe' the negative energy and related effects [178] and even to release energy from the vacuum, as seen for example in Refs. [180, 181] and references in the above-mentioned reviews in Refs. [92, 177]. In [182] by investigating the thermodynamical properties of the quantum vacuum it has been concluded that no energy can be extracted cyclically from the vacuum. Let us also note that in QED the photon propagator always remains perturbative even if it is dressed [28, 4, 183] and see also Appendix 2.A. So formally, we can define the bag constant in this theory as $B_{QED} = VED^0 - VED = -VED > 0$, since $VED^0 \equiv VED(D_0) = 0$ in the effective potential approach to leading order [156]. However, the vacuum polarization tensor is the only contribution to the photon self-energy and hence to such a defined bag constant. It has an imaginary part when photon momentum $q^2 \geq (2m)^2$, where m

is electron mass [183]. So QED vacuum is unstable against emission of electron–positron pairs from the vacuum. This means that QED vacuum, in principle, cannot have stable virtual field configurations in order to be released as the bag energy.

Since the QCD fine structure constant is strong, the idea to exploit the QCD vacuum in order to extract energy from it seems to be more attractive. Moreover, the bag constant calculated here is a manifestly gauge invariant, real and colorless (color-singlet) quantity, i.e., it can be considered as a physical quantity. In fact, here we have formulated a renormalization program to make the bag constant or, equivalently, the bag pressure finite and satisfying all other necessary requirements. The key elements of this program were the necessary subtractions at all levels. Moreover, one of its attractive features, as emphasized above, is that it is the energy density of the purely transversal severely infrared singular — thus intrinsically non-perturbative — virtual gluon field configurations which are not only stable (no imaginary part), but are in the stationary state as well, i.e., in the state with the minimum of energy as plotted on Fig. 6.5.2. That is why it makes sense to discuss the 'releasing' of the bag constant from the vacuum, more precisely, the bag energy (6.8.4). Also, the QCD vacuum will remain stable even including quark skeleton loop contributions to the bag constant, since the quark propagator cannot have an imaginary part due to the quark confinement phenomenon. It is worth emphasizing that absence of an imaginary part in the quark propagator means that it always remains an off-mass-shell object as highlighted by the first necessary condition of quark confinement formulated in Section 4.6. However, before discussing the ways to extract it, let us show that the minimum/maximum amount of energy, which can be released in a single cycle, is the bag energy, and that the QCD vacuum is an infinite source of energy.

From the quantum statistical mechanics point of view the energy is nothing but the pressure multiplied by the volume V in the infinite-volume limit [184]. So the vacuum energy, E_{vac} in terms of the bag constant is

$$E_{vac} = -B_{YM}\ V = -E_{YM}\frac{V}{\text{cm}^3} \sim -\lambda^3, \quad \text{as} \quad \lambda \to \infty, \tag{6.10.1}$$

since $V/\text{cm}^3 \sim \lambda^3$ always when the dimensionless ultraviolet cutoff λ goes to infinity. Evidently, in deriving equation (6.10.1) we use the general relation $E_{YM} = B_{YM}\ \text{cm}^3$, which is valid in any units for energy.

Let us imagine now that we can release the finite portion E_{YM} (6.8.4) from the vacuum in k different places — 'vacuum energy releasing facilities' (VERF). It can be done n_m times in each place, where $m = 1, 2, 3...k$. Then the releasing energy E_r becomes

$$E_r = E_{YM} \sum_{m=1}^{k} n_m. \tag{6.10.2}$$

The ideal case — which, however, will never be achieved — is when we could extract a finite portion of the energy an infinite number of times and in an infinite number of places. So the releasing energy (6.10.2) might be divergent as follows:

$$E_r = E_{YM} \cdot \lim_{(k,n_m)\to\infty} \sum_{m=1}^{k} n_m \sim \lambda^2, \quad \text{as} \quad \lambda \to \infty, \tag{6.10.3}$$

since the sum over m diverges quadratically in the $\lambda \to \infty$ limit, and $k \sim \lambda$, $n_m \sim \lambda$ in this case. The difference between the vacuum energy (6.10.1) and the releasing energy, E_r given by (6.10.3) is the remaining part of the vacuum energy

$$E_R = E_{vac} - E_r = E_{vac}\left[1 + \mathcal{O}(1/\lambda)\right], \quad \text{as} \quad \lambda \to \infty, \tag{6.10.4}$$

i.e., the QCD vacuum is an infinite and permanent reservoir of energy. The situation is even 'better' if one takes into account the perturbative contributions to the vacuum energy — in this case the convergence becomes of the order $\mathcal{O}(1/\lambda^2)$ in equation (6.10.4), see our previous work in [170].

The vacuum energy is badly divergent, which is not a mathematical problem. It reflects the general reality. Vacuum is everywhere and it always exists. Moreover, it is quite possible that our Universe is a special type of vacuum-excitation due to the Big Bang. As underlined above, the vacuum is an infinite and hence a permanent source of energy. The only problem is how to release the finite portion — the bag energy (6.8.4) and whether it will be profitable or not by introducing some kind of cyclic process. However, due to the huge difference between the two numbers given by (6.8.4) and (6.8.5) such a cyclically profitable process may be realistic. *We note, 'Perpetuum mobile' does not exist, but 'perpetuum source' of energy does exist, and it is the QCD ground state.*

6.11 Conclusions

In summary, we have formulated a general method to calculate numerically the quantum part of the intrinsically non-perturbative Yang–Mills vacuum energy density — the Yang–Mills bag pressure, apart from the sign, by definition — in the covariant gauge QCD. For this purpose we have used the effective potential approach for composite operators. It has an advantage of being the vacuum energy density in the absence of external sources. The bag constant is defined as a special function of the intrinsically non-perturbative effective charge integrated out over the non-perturbative, soft momentum region, see equation (6.4.1). At this stage the bag constant is colorless (color-singlet) and depends only on the transversal, physical degrees of freedom of gauge bosons. It is also free of the perturbative contributions by its construction. This has been achieved due to the subtractions at the fundamental level as given by equation (6.3.1), as well as due to all other subtractions explicitly shown in equation (6.3.6), when the gluon degrees of freedom were to be integrated out. Thus, our expressions, equations (6.4.3), (6.4.4), and (6.5.1) are general ones in order to correctly calculate the bag constant to leading order as a function of the intrinsically non-perturbative effective charge within the effective potential approach for composite operators.

The intrinsically non-perturbative effective charge depends regularly on the mass gap, which is responsible for the large-scale structure of the QCD ground state [11, 38]. The scale-setting schemes have been chosen by the two different ways, leading, nevertheless, to a rather close numerical results for the mass gap and hence for the effective scale. The calculated bag constant in addition is: finite, positive, and it has no imaginary part — thus the vacuum is stable. It is also a manifestly gauge-invariant quantity, and thus does not explicitly depend on the gauge-fixing parameter as it is required. The separation of 'soft versus hard' gluon momenta is also exact because of the maximization/minimization procedure. It becomes possible since the effective potential given by equation (6.5.2) as a function of the constant of the integration z_c has a local maximum as in Fig. 6.5.1. This also makes it possible that in the above-mentioned scale-setting schemes either the mass gap or the effective scale is only independent, since the other one is to be determined via the relation (6.5.9). Via the scale-setting scheme (6.5.10) the effective scale is independent, while in the second scale-setting scheme (6.5.11) the mass gap is independent. It is worth emphasizing that the calculated Yang–Mills energy density, ϵ_{YM}, is the energy density of the purely transversal severely infrared singular (intrinsically non-perturbative) virtual gluon field configurations which are in stationary state — in the state with the minimum of energy, as plotted in Fig. 6.5.2.

In order to compare our numerical results with phenomenology we developed a general formalism which makes it possible to relate the bag constant to the gluon condensate in a unique and self-consistent way. For this purpose we use the trace anomaly relation without applying it to the weak coupling solution for the corresponding β-function. Our numerical results turned out to be in good agreement with phenomenological values of the gluon condensate calculated and estimated within different approaches and methods [135, 163].

It is instructive to briefly summarize our theoretical and numerical results for the bag constant in general and our specific ways. The features gathered all together below are remarkable and unique.

General properties of the bag constant:

- colorless (color-singlet);
- electrically neutral;
- transversal, thus depending only on physical degrees of freedom of gauge bosons;
- free of the perturbative contributions, 'contaminations'.

Results, depending on our confining effective charge:

- the explicit gauge invariance;
- uniqueness, thus it is free of all the types of the perturbative contributions now;
- finiteness;
- positiveness;
- no imaginary part, thus it is a stable vacuum;
- existence of the stationary state for the corresponding Yang – Mills energy density (negative pressure), see Fig. 6.5.2;
- the final dependence on the mass gap only;
- a good numerical agreement with phenomenology.

Our method can be generalized on the multi-loop skeleton contributions to the effective potential approach for composite operators, as well as to take into account the quark degrees of freedom, as plotted in Fig. 6.2.1. These terms, however, will produce numerical contributions an order of magnitude less, at least, in comparison with the leading log-loop level gluon term given by equation (6.2.1). What is necessary indeed, is to be able to extract the finite part of the intrinsically non-perturbative vacuum energy density in self-consistent and manifestly gauge-invariant ways. This is provided by our method which can thus be applied to any QCD vacuum quantum and classical models at any gauge (covariant or non-covariant). It may serve as a test of them, providing an exact criterion for the separation 'stable versus unstable' vacua. Using our method we have already shown that the vacuum of classical dual Abelian Higgs model with string and without string contributions is unstable against quantum corrections [185, 186].

6.A Appendix: The general role of ghosts

Let us begin with recalling that due to the above-mentioned normalization condition in the initial equation (6.2.1), its elaborated counterpart in equation (6.2.6) depends only on the transversal, physical component of the full gluon propagator. So there is no need for ghosts to cancel its longitudinal, unphysical component, indeed. However, it is instructive to discuss the role of ghosts in general, and to clearly show that their explicit introduction leads to the same result for the bag constant, in particular.

Following Ref. [156], the effective potential at the same log-loop order for the ghost degrees of freedom analytically is:

$$V(G) = -i \int \frac{\mathrm{d}^4 k}{(2\pi)^4} \mathrm{Tr} \left[\ln \left(G_0^{-1} G \right) - \left(G_0^{-1} G \right) + 1 \right], \tag{6.A.1}$$

where $G \equiv G(k) = i/k^2 \left(1 + b\left(k^2\right)\right)$ is the full ghost propagator, $b\left(k^2\right)$ is the ghost self-energy, and $G_0 \equiv G_0(k) = i/k^2$ is its free perturbative counterpart. Trace over color group indices is assumed. Evidently, the effective potential is normalized to $V(G_0) = 0$ in the same way as the gluon part in equation (6.2.1). Substituting these expressions into the ghost term (6.A.1) and again doing some algebra in four-dimensional Euclidean space, one formally obtains that $V(G) = \epsilon_{gh} = \int \mathrm{d}k^2 f(b(k^2))$. This, in general, divergent constant contribution should be of course, regularized in order to assign to it a mathematical meaning. So the explicit functional dependence of the ghost propagator/self-energy on its argument is not important, since within the effective potential approach to calculate the vacuum energy density it is always only constant. We have to sum up all the contributions for the pure Yang – Mills fields at the same skeleton log-loop order. The relation given by equation (6.3.4) then should look like:

$$\begin{aligned} \epsilon_g + \epsilon_{gh} = &- \frac{1}{\pi^2} \int_0^{q^2_{eff}} \mathrm{d}q^2 \, q^2 \left[\ln\left[1 + 3d^{INP}\left(q^2\right)\right] - \frac{3}{4} d^{INP}\left(q^2\right) \right] \\ &+ \epsilon_{PT} + \epsilon'_{PT} + \epsilon_{gh} \, . \end{aligned} \tag{6.A.2}$$

It is worth emphasizing that, the right-hand-side of this relation may still suffer from unphysical singularities by the integral in equation (6.3.3), defining ϵ_{PT}. The problem is that the perturbative effective charge, $d^{PT}(q^2)$, which is responsible for asymptotic freedom in QCD at large q^2 may have, in general, unphysical singularities below the scale Λ^2_{QCD}, since in the equation (6.3.3) the integration goes from zero to infinity. In addition, as mentioned above, the integral (6.3.5) defining ϵ'_{PT} may be still divergent. Thus the left-hand-side of the relation (6.A.2) is indeed a formal one. It suffers from various types of unphysical singularities which may appear in its right-hand-side. In order to get a physically meaningful expression, one has to remove the two integrals (6.3.3) and (6.3.5) from equation (6.2.6). This has to be done with the help of a ghost term by imposing the following condition of cancelation of unwanted terms in the most general form:

$$\epsilon_{PT} + \epsilon'_{PT} + \epsilon_{gh} = 0. \tag{6.A.3}$$

This condition can be always fulfilled, since it is a relation between three different (unknown in general) regularized constants. Then we can rewrite the relation (6.A.2) as follows:

$$\begin{aligned} \epsilon_{YM} &= \epsilon_g - \epsilon_{PT} - \epsilon'_{PT} \\ &= -\frac{1}{\pi^2} \int\limits_0^{q^2_{eff}} dq^2\, q^2 \left[\ln\left[1 + 3d^{INP}(q^2)\right] - \frac{3}{4} d^{INP}(q^2) \right], \end{aligned} \tag{6.A.4}$$

in complete agreement with the relation (6.3.6), and hence with the definition of the bag constant given by equation (6.4.1), as it should be. So the intrinsically non-perturbative gluon contribution to the vacuum energy density has been determined by subtracting unwanted terms by means of the ghost contribution. Evidently, the subtracted terms are of no importance, while a ghost term plays no explicit role for further consideration.

In QCD the general role of ghost degrees of freedom is to cancel all the unphysical degrees of freedom of gauge bosons, thus maintaining unitarity of the S-matrix. This is the main reason why they should be taken into account always together with gluons . This means that nothing should *explicitly* depend on them after the above-mentioned cancelation is performed. One of the main purposes of their introduction is to exclude the longitudinal, unphysical component of the gluon propagator in every order of the perturbation theory, thus going beyond it and thus being a general one, indeed. If there is no need to cancel the longitudinal component of gauge boson propagators, then they should be used to eliminate the unphysical singularities of gauge bosons below the QCD asymptotic scale — as it was described above — or some other ones which may be inevitably present in any solution/ansatz for the full gluon propagator. If one knows the ghost propagator exactly, then the above-mentioned cancelation of unphysical singularities of gauge bosons should proceed automatically, as usual in the perturbation theory calculus — if, of course, all calculations are correct. For such an exact cancelation of the longitudinal part of the gluon propagator by the free perturbative ghost propagator in lower order of the perturbation theory see, for example Ref. [3]. But if it is not known exactly or known approximately — depending on the truncation/approximation scheme — as usual in the non-perturbative calculus then nevertheless, one has to impose the corresponding condition of cancelation in order to fulfill their general role. This has been done above. Thus our subtraction scheme is in agreement with the general interpretation of ghosts to cancel all the unphysical degrees of freedom of gauge bosons. So by themselves the ghosts cannot change the true dynamics of QCD, associated with the transversal component of the full gluon propagator and described by its Lorentz structure or, equivalently, by its effective charge. Nevertheless, this does not mean that we need no ghosts at all. We need them in other sectors of QCD, for example in the quark Slavnov – Taylor identity, which contains the so-called ghost–quark scattering kernel explicitly. See remarks in Section 2.9 as well.

Concluding, whatever the solution for the full ghost propagator obtained by lattice QCD and by the analytical approach based on the corresponding Schwinger – Dyson system of equations might be, it, however, should not undermine their above-mentioned general job. See more about this in Refs. [17, 45–51, 53, 243–247] respectively, and references therein. It is worth emphasizing that by no coincidence in all the papers cited above the transversal Landau gauge has been chosen by hand from the very beginning. So there is *no and cannot be an explicit* dependence on the ghost degrees of freedom in any expressions for the physical quantities, in general, and in the expression for the bag constant, in particular. In this connection, let us remind that the confining effective charge (4.4.1) is the effective charge of the relevant gluon propagator, which becomes the purely transversal in a gauge invariant way, by construction.

Problems

Problem 6.1. *Derive the relation given by equation*(6.2.5).

Problem 6.2. *Perform the integration of the expression* (6.5.2).

Problem 6.3. *Derive the relation given by equation* (6.5.6).

Chapter 7

The Non-perturbative Analytical Equation of State for the Gluon Matter I

7.1 Introduction

The prediction of a possible existence of the quark–gluon plasma (QGP) created in the relativistic heavy ion collisions is one of the most interesting theoretical achievements of QCD at non-zero temperatures and densities. A fairly full list of the relevant pioneering papers is given in [187–189] and the present status of the investigations of the properties of QCD at finite temperature and density is described in [190].

The equation of state for the QGP has been derived analytically up to the order $g^6 \ln(1/g^2)$ by using the perturbation theory expansion for the evaluation of the corresponding thermodynamic potential term by term [184, 191]. However, the most characteristic feature of the thermal perturbation theory expansion is its non-analytical dependence on the coupling constant g^2, which means that perturbative QCD is not applicable at finite temperatures, though each term has been calculated correctly. The problem is not the poor convergence of this series [184, 191–193] but rather the fact that a radius of convergence cannot even be defined; any next calculated term can be bigger than the previous one. This is an in-principle problem which cannot be overcome. From the strictly mathematical point of view, four-dimensional QCD at non-zero temperatures effectively becomes a three-dimensional theory. At the same time, three-dimensional QCD has more severe infrared singularities [40] and its coupling constant becomes dimensional. It is as a consequence of this that the dependence becomes non-analytical when using the dimensionless coupling constant g^2. One also needs to introduce three different scales, T, gT and g^2T, where T is the temperature, in order to try to understand the dynamics of the QGP within the thermal perturbation theory QCD approach.

At present, the only practical method to investigate the problem is lattice QCD at finite temperature and baryon density, which has recently shown rapid progress as described in Refs. [87, 194–198] and references therein. However, lattice QCD is primarily aimed at obtaining well-defined corresponding expressions in order to get realistic numbers for physical quantities. One may therefore get numbers and curves without understanding what the physics behind them is. Such an understanding can only come from the dynamical theory, which is continuous QCD. For example, any description of the QGP has to be formulated within the framework of a dynamical theory. The need for an analytical equation of state remains, but, of course, it should be essentially non-perturbative, approaching the so-called Stefan–Boltzmann (SB) limit only at very high temperatures. Thus the approaches of analytic non-perturbative QCD and lattice QCD to finite-temperature QCD do not exclude each other, but, on the contrary, should be complementary. This is especially true at low temperatures where the thermal QCD lattice calculations suffer from big uncertainties. On the other hand, any analytic non-perturbative approach has to correctly reproduce thermal lattice QCD results at high temperatures. There already exist interesting analytical models based on quasi-particle picture [199–208] to analyze results of $SU(3)$ lattice QCD calculations for the QGP equation of state.

The main purpose of this Chapter is to derive the non-perturbative analytical equation of state for the gluon matter of a system consisting purely of Yang–Mills fields without quark degrees of freedom. The formalism we use to generalize it to non-zero temperatures is the effective potential approach for composite operators [156]. In the absence of external sources it is nothing but the vacuum energy density. The approach is non-perturbative from the very beginning, since it deals with the expansion of the corresponding skeleton vacuum loop diagrams. The key element is the extension of our initial work [61] to non-zero temperatures, which was just the main subject of the previous Chapter. This makes it possible to introduce the temperature-dependent bag constant (pressure) as a function of the mass gap. It is this which is responsible for the large-scale structure of the QCD ground state. The confining dynamics in the gluon matter will therefore be nontrivially taken into account directly through the mass gap and via the temperature-dependent bag constant itself, but other non-perturbative effects will be also present. Let us note that the temperature-dependent bag constant within the thermal perturbative QCD has been introduced into the Gibbs equilibrium criteria for a phase transition in Refs. [199] and [209].

The effective potential approach has already been used in order to study the structure of QCD at very large baryon density for an arbitrary number of flavors [210].

7.2 The gluon pressure at zero temperature

In the previous Chapter, we have already derived an expression for the bag constant (6.4.1). Adding the bag constant to the both sides of equation (6.2.6), on account of the replacement $d \to \alpha$, and introducing the gluon pressure $P_g = \epsilon_g + B_{YM}$, one obtains

$$P_g = B_{YM} - 16 \int \frac{d^4 q}{(2\pi)^4} \left[\ln \left[1 + 3\alpha \left(q^2 \right) \right] - \frac{3}{4} \alpha(q^2) + a \right]. \qquad (7.2.1)$$

The next step is to establish the relation between the full effective charge $\alpha(q^2)$ and its intrinsically non-perturbative counterpart $\alpha^{INP}(q^2)$, which has already been done in Part I and was repeated in the previous Chapter, namely $\alpha^{INP}(q^2; \Delta_R^2) = \alpha(q^2; \Delta_R^2) - \alpha(q^2; \Delta_R^2 = 0) = \alpha(q^2; \Delta_R^2) - \alpha^{PT}(q^2)$. However, this is not the whole story yet! Since the perturbative part $\alpha^{PT}(q^2)$ contains the free gluon Lorentz structure $\alpha^0(q^2) = 1$, it should be directly extracted as follows:

$$\alpha \left(q^2 \right) = \alpha^{INP} \left(q^2 \right) + \alpha^{PT} \left(q^2 \right) = \alpha^{INP} \left(q^2 \right) + 1 + \alpha^{AF} \left(q^2 \right), \qquad (7.2.2)$$

where we have dropped the explicit dependence on Δ_R^2, and $\alpha^{AF}(q^2)$ describes the part responsible for asymptotic freedom in the perturbation theory gluon effective charge. This procedure is necessary in order to maintain the normalization of the free perturbation theory vacuum to zero. In this connection, let us stress that the second equality in this relation does not imply any violation of asymptotic freedom in the full gluon effective charge $\alpha(q^2)$, provided by the perturbation theory effective charge $\alpha^{PT}(q^2)$ in the first equality of the same relation. Extracting $\alpha^0(q^2) = 1$ explicitly, we thereby subtract the divergent contribution associated with the constant a in equation (7.2.1), and thus the above-mentioned normalization condition will be automatically satisfied.

Substituting the decomposition (7.2.2) into equation (7.2.1), the lengthy algebra leads to

$$P_g = P_{NP} + P_{PT} = B_{YM} + P_{YM} + P_{PT}, \qquad (7.2.3)$$

where

$$B_{YM} = 16 \int^{q^2_{eff}} \frac{d^4q}{(2\pi)^4} \left[\ln \left[1 + 3\alpha^{INP}(q^2)\right] - \frac{3}{4}\alpha^{INP}(q^2) \right], \quad (7.2.4)$$

$$P_{YM} = -16 \int \frac{d^4q}{(2\pi)^4} \left[\ln \left[1 + \frac{3}{4}\alpha^{INP}(q^2)\right] - \frac{3}{4}\alpha^{INP}(q^2) \right], \quad (7.2.5)$$

and

$$P_{PT} = -16 \int_{\Lambda^2_{YM}} \frac{d^4q}{(2\pi)^4} \left[\ln \left[1 + \frac{3\alpha^{PT}(q^2)}{4 + 3\alpha^{INP}(q^2)}\right] - \frac{3}{4}\alpha^{PT}(q^2) \right]. \quad (7.2.6)$$

Let us underline that from now on the sum of equations (7.2.4) and (7.2.5) depends solely on the intrinsically non-perturbative effective charge $\alpha^{INP}(q^2)$; nevertheless, for the simplicity, it will be denoted as the non-perturbative (NP) pressure P_{NP}, as has been already done in equation (7.2.3). In the Yang–Mills term P_{YM}, given in equation (7.2.5), the integration over variable q^2 goes from zero to infinity, while in the perturbation theory term (7.2.6) it is indicated that the integration over q^2 cannot go below Λ^2_{YM}. Let us also note that after extracting the free gluon effective charge described above, we again denote the remaining perturbation theory part as $\alpha^{PT}(q^2)$ (see remarks below).

Let us remind that the intrinsically non-perturbative effective charge is

$$\alpha^{INP}(q^2) = \frac{\Delta^2_R}{q^2}, \quad (7.2.7)$$

where $\Delta^2_R \equiv \Delta^2_{JW}$ is the Jaffe–Witten (JW) mass gap [11], which is responsible for the large-scale structure of the QCD vacuum, and thus for its intrinsically non-perturbative dynamics. How the regularized mass gap appears in QCD without the violation of its exact $SU(3)$ color gauge invariance has been described in Part I of this book, where its non-perturbative multiplicative renormalization program has also been performed.

The P_{PT} part, shown in equation (7.2.6), together with the confining effective charge (7.2.7) depends mainly on the perturbation theory effective charge $\alpha^{PT}(q^2)$. It has been derived in Part I and is given by equation (4.9.4), namely

$$\alpha^{PT}(q^2) = \frac{\alpha_s}{1 + \alpha_s b \ln(q^2/\Lambda^2_{YM})}. \quad (7.2.8)$$

Here, $\Lambda^2_{YM} = 0.09$ GeV^2 [211] is the asymptotic scale parameter for $SU(3)$ Yang – Mills and $b = (11/4\pi)$ for these fields, while the strong fine-structure constant is $\alpha_s = \alpha_s(m_Z) = 0.1184$ [167]. In equation (7.2.8) q^2 cannot go below Λ^2_{YM}, i.e., $\Lambda^2_{YM} \leq q^2 \leq \infty$, which has already been symbolically shown in equation (7.2.6). As mentioned above, we have left the notation $\alpha^{AF}(q^2)$ for the asymptotic freedom relation $\alpha^{AF}(q^2) = 1/b\ln(q^2/\Lambda^2_{YM})$ itself. One can recover it from equation (7.2.8) in the $q^2 \to \infty$ limit. The interaction is formally switched off by letting $\alpha^{PT}(q^2) = \alpha_s^{INP}(q^2) = 0$, or, equivalently, $\alpha_s = \Delta^2_R = 0$, then $P_{NP} = P_{PT} = 0$, so that $P_g = 0$ as well. This is due to the initial normalization condition of the free perturbative vacuum to zero in the effective potential approach up to the leading skeleton log-loop order (see the previous Chapter).

7.3 The gluon pressure at non-zero temperature

In the imaginary-time formalism introduced in Refs. [184, 212, 213], all of the four-dimensional integrals can be easily generalized to non-zero temperature T according to the prescription (let us remind that the signature is Euclidean from the very beginning)

$$\int \frac{dq_0}{(2\pi)} \to T \sum_{n=-\infty}^{+\infty}, \quad q^2 = \mathbf{q}^2 + q_0^2 = \mathbf{q}^2 + \omega_n^2 = \omega^2 + \omega_n^2, \quad \omega_n = 2n\pi T. \tag{7.3.1}$$

In other words, each integral over q_0 of the loop momentum is to be replaced by the sum over the Matsubara frequencies labeled by n, which obviously assumes the replacement $q_0 \to \omega_n = 2n\pi T$ for bosons (gluons). In frequency–momentum space the effective charges (7.2.7) and (7.2.8) become

$$\alpha^{INP}\left(q^2\right) = \alpha^{INP}\left(\mathbf{q}^2, \omega_n^2\right) = \alpha^{INP}\left(\omega^2, \omega_n^2\right) = \frac{\Delta_R^2}{\omega^2 + \omega_n^2}, \tag{7.3.2}$$

and

$$\alpha^{PT}\left(q^2\right) = \alpha^{PT}\left(\mathbf{q}^2, \omega_n^2\right) = \alpha^{PT}\left(\omega^2, \omega_n^2\right) = \frac{\alpha_s}{1 + \alpha_s b \ln\left(\omega^2 + \omega_n^2/\Lambda^2_{YM}\right)}, \tag{7.3.3}$$

respectively. It is also convenient to introduce the following notations:

$$T^{-1} = \beta \text{ and } \omega = \sqrt{\mathbf{q}^2}, \tag{7.3.4}$$

where, evidently, in all the expressions $\mathbf{q}^2$ is the square of the three-dimensional loop momentum, in complete agreement with the relations (7.3.1).

Introducing the temperature dependence for both sides of the relation (7.2.3), we obtain

$$P_g(T) = P_{NP}(T) + P_{PT}(T) = B_{YM}(T) + P_{YM}(T) + P_{PT}(T), \quad (7.3.5)$$

where the corresponding terms in frequency–momentum space are:

$$B_{YM}(T) = \frac{8}{\pi^2} \int_0^{\omega_{eff}} d\omega \, \omega^2 \, T \times \sum_{n=-\infty}^{+\infty} \left[\ln \left[1 + 3\alpha^{INP}\left(\omega^2, \omega_n^2\right)\right] - \frac{3}{4}\alpha^{INP}\left(\omega^2, \omega_n^2\right) \right], \quad (7.3.6)$$

$$P_{YM}(T) = -\frac{8}{\pi^2} \int_0^{\infty} d\omega \, \omega^2 \, T \times \sum_{n=-\infty}^{+\infty} \left[\ln \left[1 + \frac{3}{4}\alpha^{INP}\left(\omega^2, \omega_n^2\right)\right] - \frac{3}{4}\alpha^{INP}\left(\omega^2, \omega_n^2\right) \right], \quad (7.3.7)$$

$$P_{PT}(T) = -\frac{8}{\pi^2} \int_{\Lambda_{YM}}^{\infty} d\omega \, \omega^2 \, T \times \sum_{n=-\infty}^{+\infty} \left[\ln \left[1 + \frac{3\alpha^{PT}\left(\omega^2, \omega_n^2\right)}{4 + 3\alpha^{INP}\left(\omega^2, \omega_n^2\right)}\right] - \frac{3}{4}\alpha^{PT}\left(\omega^2, \omega_n^2\right) \right]. \quad (7.3.8)$$

The non-perturbative pressure $P_{NP}(T) = B_{YM}(T) + P_{YM}(T)$ and the perturbative pressure $P_{PT}(T)$, and hence the gluon pressure $P_g(T)$ given by equation (7.3.7), are normalized to zero when the interaction is formally switched off, via setting $\alpha_s = \Delta_R^2 = 0$. This means that the initial normalization condition of the free perturbative vacuum to zero holds at non-zero temperature as well.

7.4 The scale-setting scheme

Let us note that ω_{eff}, which appears first in the integral (7.3.6), is the only free parameter of our approach. In frequency–momentum space it is

$$\omega_{eff} = \sqrt{q_{eff}^2 - \omega_c^2}, \tag{7.4.1}$$

where we introduce the constant Matsubara frequency ω_c, which is always positive. Hence ω_{eff} is always less or equal to the q_{eff} of four-dimensional QCD, so $\omega_{eff} \leq q_{eff}$. One can then conclude that q_{eff} is a very good upper limit for possible values of ω_{eff}. In this connection, let us recall that the bag constant B_{YM} at zero temperature has been successfully calculated at a scale $q_{eff}^2 = 1$ GeV^2, in fair agreement with other phenomenological quantities such as gluon condensate as described in the previous Chapter. So ω_{eff} is fixed as follows:

$$\omega_{eff} = \sqrt{q_{eff}^2} = 1 \text{ GeV}. \tag{7.4.2}$$

The mass gap squared Δ_R^2, also calculated at this scale, has a value

$$\Delta_R^2 = 0.4564 \text{ GeV}^2, \quad \text{and thus} \quad \Delta_R = 0.6756 \text{ GeV}. \tag{7.4.3}$$

The effective gluon masses, defined in the relations (7.5.3) and (7.5.8) below, then numerically become

$$m'_{eff} = \sqrt{3}\Delta_R = 1.17 \text{ GeV}, \quad \bar{m}_{eff} = \frac{\sqrt{3}}{2}\Delta_R = 0.585 \text{ GeV}. \tag{7.4.4}$$

7.5 The $P_{NP}(T)$ contribution

One of the attractive features of the confining effective charge (7.3.2) is that it allows an exact summation over the Matsubara frequencies in the non-perturbative pressure

$$P_{NP}(T) = B_{YM}(T) + P_{YM}(T), \tag{7.5.1}$$

given by the sum of the integrals (7.3.6) and (7.3.7).

Derivation of $B_{YM}(T)$

It is convenient to begin with equation (7.3.6) for the bag constant. After the substitution of the expression from equation (7.3.2), in frequency–momentum one obtains

$$B_{YM}(T) = 16 \int \frac{d^3 q}{(2\pi)^3} \, T \sum_{n=-\infty}^{+\infty} \left[\ln \left[\frac{\omega'^2 + \omega_n^2}{\omega^2 + \omega_n^2} \right] - \frac{3}{4} \frac{\Delta_R^2}{\omega^2 + \omega_n^2} \right], \quad (7.5.2)$$

where we introduced the notations:

$$\omega' = \sqrt{\omega^2 + m_{eff}'^2}, \quad \text{and} \quad m_{eff}' = \sqrt{3}\Delta_R. \quad (7.5.3)$$

Substituting all the results of the summation over the Matsubara frequencies — see Appendix 7.A — for the equation (7.5.2), dropping the β-independent terms based on Ref. [184], and performing almost trivial integration over angular variables, one gets

$$B_{YM}(T) = -\frac{8}{\pi^2} \int_0^{\omega_{eff}} d\omega \, \omega^2 \left[\frac{3}{4} \Delta_R^2 \frac{1}{\omega} \frac{1}{e^{\beta\omega} - 1} - 2\beta^{-1} \ln \left[\frac{1 - e^{-\beta\omega'}}{1 - e^{-\beta\omega}} \right] \right]. \quad (7.5.4)$$

It is convenient to present the integral (7.24) as a sum of several terms

$$B_{YM}(T) = -\frac{6}{\pi^2} \Delta_R^2 B_{YM}^{(1)}(T) - \frac{16}{\pi^2} T \left[B_{YM}^{(2)}(T) - B_{YM}^{(3)}(T) \right], \quad (7.5.5)$$

where the explicit expressions of all the terms are given as follows:

$$B_{YM}^{(1)}(T) = \int_0^{\omega_{eff}} d\omega \frac{\omega}{e^{\beta\omega} - 1},$$

$$B_{YM}^{(2)}(T) = \int_0^{\omega_{eff}} d\omega \, \omega^2 \ln \left[1 - e^{-\beta\omega} \right],$$

$$B_{YM}^{(3)}(T) = \int_0^{\omega_{eff}} d\omega \, \omega^2 \ln \left[1 - e^{-\beta\omega'} \right]. \quad (7.5.6)$$

Derivation of $P_{YM}(T)$

Let us continue with equation (7.3.7) for the Yang–Mills part $P_{YM}(T)$. After substituting the expression (7.3.3), in frequency–momentum space one obtains

$$P_{YM}(T) = -16 \int \frac{d^3q}{(2\pi)^3}\, T \sum_{n=-\infty}^{+\infty} \left[\ln \left[\frac{\bar{\omega}^2 + \omega_n^2}{\omega^2 + \omega_n^2} \right] - \frac{3}{4} \frac{\Delta_R^2}{\omega^2 + \omega_n^2} \right], \quad (7.5.7)$$

where we introduced the following notations:

$$\bar{\omega} = \sqrt{\omega^2 + \bar{m}_{eff}^2} \quad \text{and} \quad \bar{m}_{eff} = \frac{\sqrt{3}}{2}\Delta_R = \frac{1}{2} m'_{eff}. \quad (7.5.8)$$

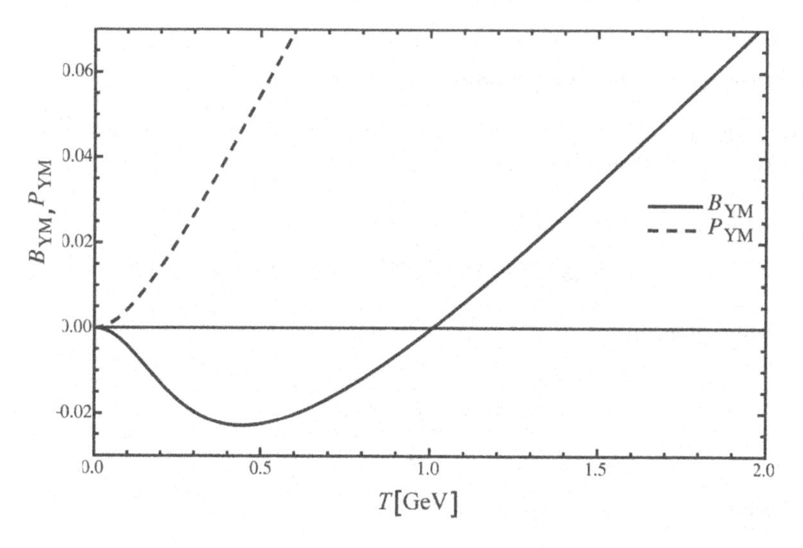

Fig. 7.5.1 The bag pressure (7.5.5) and the non-perturbative Yang–Mills pressure (7.5.9) in GeV^4 units as functions of temperature T in GeV units. The bag constant at $T = 1$ GeV is zero, so up to this value it is responsible for the non-perturbative vacuum contributions to the pressure (7.5.11).

Comparing equations (7.5.7) and (7.5.2) one can write down the final result directly. For this purpose, in the system of equations (7.5.5) and (7.5.6) one must change the overall sign, replace ω' by $\bar{\omega}$ and integrate from zero to infinity. Thus, one obtains

$$P_{YM}(T) = \frac{6}{\pi^2} \Delta_R^2 P_{YM}^{(1)}(T) + \frac{16}{\pi^2} T \left[P_{YM}^{(2)}(T) - P_{YM}^{(3)}(T) \right], \quad (7.5.9)$$

where

$$P^{(1)}_{YM}(T) = \int_0^\infty d\omega \frac{\omega}{e^{\beta\omega} - 1} = \frac{\pi^2}{6} T^2,$$

$$P^{(2)}_{YM}(T) = \int_0^\infty d\omega\, \omega^2 \ln\left[1 - e^{-\beta\omega}\right],$$

$$P^{(3)}_{YM}(T) = \int_0^\infty d\omega\, \omega^2 \ln\left[1 - e^{-\beta\bar{\omega}}\right]. \qquad (7.5.10)$$

The bag pressure (7.5.5) and the Yang–Mills part (7.5.9) in GeV^4 units as functions of temperature T are shown in Fig. 7.5.1.

The non-perturbative pressure

Summing up all the expressions and integrals (7.5.5)–(7.5.6) and (7.5.9)–(7.5.10), one obtains

$$\begin{aligned} P_{NP}(T) &= B_{YM}(T) + P_{YM}(T) \\ &= \frac{6}{\pi^2}\Delta_R^2 P_1(T) + \frac{16}{\pi^2} T\left[P_2(T) + P_3(T) - P_4(T)\right], \end{aligned} \qquad (7.5.11)$$

in which equation

$$P_1(T) = \int_{\omega_{eff}}^\infty d\omega \frac{\omega}{e^{\beta\omega} - 1}, \qquad (7.5.12)$$

and while

$$P_2(T) = \int_{\omega_{eff}}^\infty d\omega\, \omega^2 \ln\left[1 - e^{-\beta\omega}\right],$$

$$P_3(T) = \int_0^{\omega_{eff}} d\omega\, \omega^2 \ln\left[1 - e^{-\beta\omega'}\right],$$

$$P_4(T) = \int_0^\infty d\omega\, \omega^2 \ln\left[1 - e^{-\beta\bar{\omega}}\right]. \qquad (7.5.13)$$

Let us recall once more that in all these integrals $\beta = T^{-1}$, ω_{eff} along with the mass gap Δ_R^2 are fixed, while ω' and $\bar{\omega}$ are given by the relations (7.5.3) and (7.5.8), respectively. In the formal perturbation theory $\Delta_R^2 = 0$ limit it follows that $\bar{\omega} = \omega' = \omega$ and the combination $P_2(T) + P_3(T) - P_4(T)$ becomes identically zero. Thus the non-perturbative part (7.5.11) of the gluon pressure (7.3.5) in this limit vanishes.

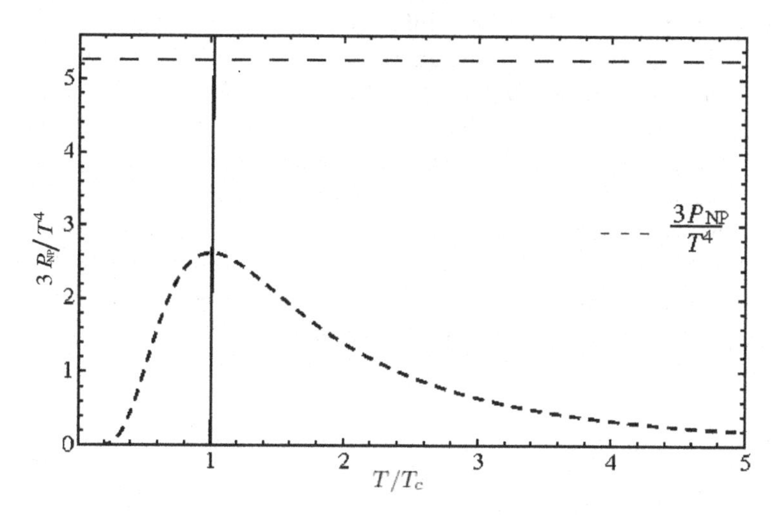

Fig. 7.5.2 The non-perturbative pressure (7.5.11) properly scaled is shown as a function of T/T_c. It has a maximum at $T_c = 266.5$ MeV. Here the horizontal dashed line is the general Stefan–Boltzmann constant $3P_{SB}(T)/T^4 = 24\pi^2/45 \approx 5.2634$.

7.6 Conclusions

The non-perturbative pressure as a function of temperature is shown in Fig. 7.5.2. It has a maximum at some finite characteristic temperature, $T = T_c = 266.5$ MeV. Confining dynamics in the non-perturbative pressure (7.5.11) is implemented via the bag pressure (7.5.5) and the mass gap itself. However, other non-perturbative effects are also present via $P_{YM}(T)$ and $P_{PT}(T)$, given in equation (7.5.9) and (7.3.8), respectively, though in equation (7.3.8) they are logarithmically suppressed. One of the interesting numerical features of our calculations is:

$$[P_{SB}(T) - 2P_{NP}(T)]_{T=T_c} = 0,$$
$$\frac{\partial}{\partial T}[P_{SB}(T) - 2P_{NP}(T)]_{T=T_c} = 0, \tag{7.6.1}$$

where, let us note in advance, the Stefan – Boltzmann pressure is: $P_{SB}(T) = (8\pi^2/45)T^4$.

However, the most interesting physical features of our calculations are:

(i) The mass gap Δ_R^2 or, equivalently, ω_{eff} is the only independent input scale parameter needed to calculate the non-perturbative pressure.

(ii) The presence of the two different types of massive gluonic excitations ω' and $\bar{\omega}$, given in equations (7.5.3) and (7.5.8) with effective masses 1.17 GeV and 0.585 GeV, respectively.

(iii) We also have the two different types of massless gluonic excitations ω_1 and ω_2 which, nevertheless, are not free. They propagate in accordance with equation (7.5.12) and the first of integrals (7.5.13), multiplied by the factors $6\Delta_R^2/\pi^2$ and $16T/\pi^2$, respectively.

(iv) All these massive and massless gluonic excitations are of non-perturbative dynamical origin. They disappear from the gluon matter spectrum in the perturbation theory $\Delta_R^2 = 0$ limit.

7.A Appendix: The summation of the thermal logarithms

In the second terms of equations (7.5.2) and (7.5.7) the summation over the Matsubara frequencies can be done explicitly following Ref. [184]:

$$\begin{aligned}\sum_{n=-\infty}^{+\infty} \frac{1}{\omega^2+\omega_n^2} &= \sum_{n=-\infty}^{\infty} \frac{1}{\omega^2+(2\pi T)^2 n^2} \\ &= \left(\frac{\beta}{2\pi}\right)^2 \sum_{n=-\infty}^{+\infty} \frac{1}{n^2+(\beta\omega/2\pi)^2} \\ &= \left(\frac{\beta}{2\pi}\right)^2 \frac{2\pi^2}{\beta\omega}\left(1+\frac{2}{e^{\beta\omega}-1}\right) \\ &= \frac{\beta}{2\omega}\left(1+\frac{2}{e^{\beta\omega}-1}\right). \end{aligned} \tag{7.A.1}$$

In terms of the above-introduced parameters, the sums in equation (7.5.2) containing the corresponding logarithms look like:

$$\sum_{n=-\infty}^{+\infty} \ln\left[3\Delta^2+\omega^2+\omega_n^2\right] = \ln\omega'^2 + 2\sum_{n=1}^{\infty} \ln\left(\frac{2\pi}{\beta}\right)^2 \left[n^2+(\beta\omega'/2\pi)^2\right] \tag{7.A.2}$$

and

$$\sum_{n=-\infty}^{+\infty} \ln\left[\omega^2+\omega_n^2\right] = \ln\omega^2 + 2\sum_{n=1}^{\infty} \ln\left(\frac{2\pi}{\beta}\right)^2 \left[n^2+(\beta\omega/2\pi)^2\right]. \tag{7.A.3}$$

It is convenient to introduce the notations:

$$L(\omega') = \sum_{n=1}^{\infty} \ln\left[n^2+(\beta\omega'/2\pi)^2\right] = \sum_{n=1}^{\infty} \ln n^2 + \sum_{n=1}^{\infty} \ln\left[1-\frac{x'^2}{n^2\pi^2}\right] \tag{7.A.4}$$

and similarly, letting $\omega' \to \bar{\omega}$

$$L(\omega) = \sum_{n=1}^{\infty} \ln[n^2+(\beta\omega/2\pi)^2] = \sum_{n=1}^{\infty} \ln n^2 + \sum_{n=1}^{\infty} \ln\left[1-\frac{x^2}{n^2\pi^2}\right]. \tag{7.A.5}$$

In these expressions we introduced the following notations:

$$x'^2 = -\left(\frac{\beta\omega'}{2}\right)^2, \quad \text{and} \quad x^2 = -\left(\frac{\beta\omega}{2}\right)^2. \tag{7.A.6}$$

So the difference $L(\omega') - L(\omega)$ becomes

$$L(\omega') - L(\omega) = \sum_{n=1}^{\infty} \ln\left[1 - \frac{x'^2}{n^2\pi^2}\right] - \sum_{n=1}^{\infty} \ln\left[1 - \frac{x^2}{n^2\pi^2}\right]$$
$$= \ln\sin x' - \frac{1}{2}\ln x'^2 - \ln\sin x + \frac{1}{2}\ln x^2, \tag{7.A.7}$$

or, equivalently,

$$L(\omega') - L(\omega) = -\frac{1}{2}\ln\left(\frac{x'^2}{x^2}\right) + \ln\left(\frac{\sin x'}{\sin x}\right). \tag{7.A.8}$$

From the relation (7.A.6) it follows that

$$x' = \pm i\left(\frac{\beta\omega'}{2}\right), \quad \text{and} \quad x = \pm i\left(\frac{\beta\omega}{2}\right), \tag{7.A.9}$$

so equation (7.A.8) finally becomes

$$L(\omega') - L(\omega) = -\frac{1}{2}\ln\left(\frac{\omega'^2}{\omega^2}\right) + \frac{1}{2}\beta(\omega' - \omega) + \ln\left(\frac{1 - e^{-\beta\omega'}}{1 - e^{-\beta\omega}}\right). \tag{7.A.10}$$

Problems

Problem 7.1. *Using the decomposition* (7.2.2), *derive the system of equations* (7.2.4)–(7.2.6) *from equation* (7.2.1).

Problem 7.2. *Perform the summation of the thermal logarithms described in Appendix 7.A.*

Problem 7.3. *Derive the equation* (7.5.4).

Problem 7.4. *Derive the equation* (7.5.5)–(7.5.6) *from equation* (7.5.4).

Problem 7.5. *Derive equations* (7.5.9)–(7.5.10).

Problem 7.6. *Derive equations* (7.5.11)–(7.5.13).

Chapter 8

The Non-perturbative Analytical Equation of State for the Gluon Matter II

8.1 Introduction

The main purpose of this Chapter is to continue the derivation of the analytical equation of state for the gluon matter — a system consisting purely of $SU(3)$ Yang–Mills fields without quark degrees of freedom. Its non-perturbative part which solely depends on the mass gap has been evaluated in the previous Chapter. The perturbative part given by equation (7.3.8) together with the mass gap depends on the QCD fine structure constant α_s. Here we are going to formulate and develop the analytic formalism which makes it possible to determine the perturbative part of the Yang–Mills equation of state or, equivalently, the gluon matter equation of state in terms of the convergent series in integer powers of α_s. So this allows us to calculate the perturbative contributions termwise in all orders of a small α_s. We will also explicitly derive and numerically calculate the first perturbative contribution of the α_s-order to the non-perturbative part of the gluon matter equation of state derived and calculated previously. The low- and high-temperature expansions for the gluon pressure will be analytically evaluated as well.

8.2 Analytic thermal perturbation theory

Let us begin here reminding that we were able to perform the summation over the Matsubara frequencies analytically (exactly) for the non-perturbative part of the gluon pressure (7.3.5). To do the same for its perturbative part (7.3.8) is a formidable task. The only way to evaluate it is the numerical summation over n and the integration over ω, which is beyond our capabilities at present — if it is possible at all. Our primary goal in this chapter is threefold. Firstly, to develop the analytical formalism which makes it possible to calculate $P_{PT}(T)$ (7.3.8) termwise in integer powers of a small α_s. Secondly, to calculate explicitly the perturbative contribution of the α_s-order. Thirdly, to derive the low- and high-temperature expansions for the gluon pressure.

For the first goal, it is convenient to re-write the integral (7.3.8) as follows:

$$P_{PT}(T) = -\frac{8}{\pi^2} \int_{\Lambda_{YM}}^{\infty} d\omega\, \omega^2\, T \sum_{n=-\infty}^{+\infty} \left[\ln\left[1 + x\left(\omega^2, \omega_n^2\right)\right] - \frac{3}{4}\alpha^{PT}\left(\omega^2, \omega_n^2\right) \right], \tag{8.2.1}$$

where

$$x\left(\omega^2, \omega_n^2\right) = \frac{3\alpha^{PT}\left(\omega^2, \omega_n^2\right)}{4 + 3\alpha^{INP}\left(\omega^2, \omega_n^2\right)} = \frac{3}{4}\frac{\left(\omega^2 + \omega_n^2\right)}{M\left(\bar{\omega}^2, \omega_n^2\right)} \frac{\alpha_s}{1 + \alpha_s \ln z_n}, \tag{8.2.2}$$

with the help of the expressions (7.3.2) and (7.3.3). Here

$$\begin{aligned} M\left(\bar{\omega}^2, \omega_n^2\right) &= \bar{\omega}^2 + \omega_n^2, \\ \ln z_n \equiv \ln z\left(\omega^2, \omega_n^2\right) &= b \ln\left[\left(\omega^2 + \omega_n^2\right) / \Lambda_{YM}^2\right], \end{aligned} \tag{8.2.3}$$

and $\bar{\omega}^2$ is given in equation (7.5.8). Let us also note that in these notations

$$\alpha^{PT}\left(\omega^2, \omega_n^2\right) \equiv \alpha(z_n) = \frac{\alpha_s}{1 + \alpha_s \ln z_n}. \tag{8.2.4}$$

There is an interesting observation concerning the argument $x\left(\omega^2, \omega_n^2\right)$ of the logarithm, $\ln\left[1 + x(\omega^2, \omega_n^2)\right]$ in the integral (8.2.1). At its lower limit $\omega = \Lambda_{YM}$ and $n = 0$ the argument (8.2.2) numerically becomes

$$x\left(\Lambda_{YM}^2\right) = \frac{3\alpha_s \Lambda_{YM}^2}{4\Lambda_{YM}^2 + 3\Delta_R^2} = 0.0185, \tag{8.2.5}$$

and the numerical values of Δ_R^2, α_s and Λ_{YM}^2 given in the previous Chapters have already been used. The argument of the logarithm is really small (it is an order of magnitude smaller than α_s itself), and it will become even smaller and smaller with ω and n going to infinity. This means that it is legitimate to expand the logarithm $\ln\left[1 + x\left(\omega^2, \omega_n^2\right)\right]$ in the integral (8.2.1) in powers of small $x\left(\omega^2,\omega_n^2\right)$ at any n and in the whole range of the integration over ω, i.e. $\infty \geq \omega \geq \Lambda_{YM}$. Doing so, following Ref. [214] one obtains

$$\ln\left[1+x\left(\omega^2,\omega_n^2\right)\right] = -\sum_{m=1}^{\infty}\frac{(-1)^m}{m}x^m\left(\omega^2,\omega_n^2\right), \text{ where } x(\omega^2,\omega_n^2) \ll 1. \tag{8.2.6}$$

Extracting the first term in expansion(8.2.6) the integral (8.2.1) becomes

$$P_{PT}(T) = -\frac{8}{\pi^2}\int_{\Lambda_{YM}}^{\infty} d\omega\, \omega^2\, T \times \sum_{n=-\infty}^{+\infty}\left[\left[x\left(\omega^2,\omega_n^2\right)-\frac{3}{4}\alpha(z_n)\right]-\sum_{m=2}^{\infty}\frac{(-1)^m}{m}x^m\left(\omega^2,\omega_n^2\right)\right]. \tag{8.2.7}$$

From now on it is instructive to separate the two terms in the integral (8.2.7) as follows:

$$P_{PT}(T) = P_{PT}\left(\Delta_R^2;T\right)+P'_{PT}(T), \tag{8.2.8}$$

where

$$P_{PT}\left(\Delta_R^2;T\right) = -\frac{8}{\pi^2}\int_{\Lambda_{YM}}^{\infty} d\omega\, \omega^2\, T\sum_{n=-\infty}^{+\infty}\left[x\left(\omega^2,\omega_n^2\right)-(3/4)\,\alpha\left(z_n\right)\right]$$
$$= \frac{9}{2\pi^2}\Delta_R^2\int_{\Lambda_{YM}}^{\infty} d\omega\, \omega^2\, T\sum_{n=-\infty}^{+\infty}\left[\frac{1}{M\left(\bar{\omega}^2,\omega_n^2\right)}\frac{\alpha_s}{1+\alpha_s\ln z_n}\right], \tag{8.2.9}$$

on account of the relations (8.2.2)–(8.2.4), and with $x\left(\omega^2,\omega_n^2\right) \ll 1$,

$$P'_{PT}(T) = \frac{8}{\pi^2}\int_{\Lambda_{YM}}^{\infty} d\omega\, \omega^2\, T\sum_{n=-\infty}^{+\infty}\left[\sum_{m=2}^{\infty}\frac{(-1)^m}{m}x^m\left(\omega^2,\omega_n^2\right)\right]. \tag{8.2.10}$$

The $P_{PT}(\Delta_R^2;T)$ contribution

The principal difference between the two terms given by equations (8.2.9) and (8.2.10) is that the former term vanishes in the formal $\Delta_R^2 = 0$ limit, while the latter one survives it. Let us consider the first term in more detail. The function $(1+\alpha_s \ln z_n)^{-1}$ in the integral (8.2.9) can formally be replaced by the expansion in integer powers of α_s, namely

$$(1+\alpha_s \ln z_n)^{-1} = \sum_{k=0}^{\infty} (-1)^k \alpha_s^k \ln^k z_n, \tag{8.2.11}$$

which converges to the corresponding function in the whole range $\Lambda_{YM} \leq \omega \leq \infty$ and at any n for $|\alpha_s \ln z_n| < 1$. The integral (8.9) can equivalently be re-written as follows:

$$P_{PT}(\Delta_R^2;T) = \frac{9}{2\pi^2}\Delta_R^2 \alpha_s \int_{\Lambda_{YM}}^{\infty} d\omega\, \omega^2\, T \times \sum_{n=-\infty}^{+\infty} \left[\frac{1}{M(\bar{\omega}^2, \omega_n^2)} \sum_{k=0}^{\infty} (-1)^k \alpha_s^k \ln^k z_n \right], \tag{8.2.12}$$

which makes it possible to present it as a sum in integer powers of α_s, namely

$$P_{PT}\left(\Delta_R^2;T\right) = \sum_{k=1}^{\infty} \alpha_s^k P_k\left(\Delta_R^2;T\right), \tag{8.2.13}$$

where

$$P_k\left(\Delta_R^2;T\right) = \frac{9}{2\pi^2}\Delta_R^2 \int_{\Lambda_{YM}}^{\infty} d\omega\, \omega^2\, T \sum_{n=-\infty}^{+\infty} \left[\frac{1}{M(\bar{\omega}^2, \omega_n^2)} (-1)^{k-1} \ln^{k-1} z_n \right]. \tag{8.2.14}$$

For example, the first term $P_1\left(\Delta_R^2;T\right)$ explicitly looks like

$$P_1\left(\Delta_R^2;T\right) = \frac{9}{2\pi^2}\Delta_R^2 \int_{\Lambda_{YM}}^{\infty} d\omega\, \omega^2\, T \sum_{n=-\infty}^{+\infty} \frac{1}{M(\bar{\omega}^2, \omega_n^2)}, \tag{8.2.15}$$

where $M(\bar{\omega}^2, \omega_n^2) = \bar{\omega}^2 + \omega_n^2$, and $\bar{\omega}^2$ itself is given in the relation (7.5.8). In this integral the summation over the Matsubara frequencies can be performed analytically (exactly) with the help of the expression explicitly given in Appendix 7.A. Omitting all the derivation and dropping the β-independent terms, following Ref. [184], one obtains

$$P_1\left(\Delta_R^2;T\right) = \frac{9}{2\pi^2}\Delta_R^2 \int\limits_{\Lambda_{YM}}^{\infty} d\omega\ \omega^2\ \frac{1}{\bar{\omega}}\frac{1}{e^{\beta\bar{\omega}}-1}. \tag{8.2.16}$$

The $P'_{PT}(T)$ contribution

On account of the relations (8.2.2)-(8.2.4), the integral (8.2.10) becomes

$$P'_{PT}(T) = \frac{8}{\pi^2}\int\limits_{\Lambda_{YM}}^{\infty} d\omega\ \omega^2\ T\sum_{n=-\infty}^{+\infty}\left[\sum_{m=2}^{\infty} b_m\left(\omega^2,\omega_n^2\right)\frac{\alpha_s^m}{\left(1+\alpha_s\ln z_n\right)^m}\right], \tag{8.2.17}$$

where

$$b_m\left(\omega^2,\omega_n^2\right) = -\frac{(-1)^m 3^m}{m4^m}\frac{\left(\omega^2+\omega_n^2\right)^m}{M^m\left(\bar{\omega}^2,\omega_n^2\right)}. \tag{8.2.18}$$

In complete analogy with the expansion (8.2.11) one gets [214]

$$\left(1+\alpha_s\ln z_n\right)^{-m} = \sum_{k=0}^{\infty} c_k(m)\alpha_s^k \ln^k z_n, \tag{8.2.19}$$

where

$$c_0(m) = 1, \text{ and } c_p(m) = \frac{1}{p}\sum_{k=1}^{p}(km-p+k)(-1)^k c_{p-k}, \text{ with } p \geq 1. \tag{8.2.20}$$

What has been said in connection with the expansion (8.2.11) is valid for the expansion (8.2.19) as well. So on its account, the integral (8.2.17) can equivalently be re-written as follows:

$$P'_{PT}(T) = \frac{8}{\pi^2}\int\limits_{\Lambda_{YM}}^{\infty} d\omega\ \omega^2\ T\sum_{n=-\infty}^{+\infty}\left[\sum_{m=2}^{\infty} b_m(\omega^2,\omega_n^2)\alpha_s^m \sum_{k=0}^{\infty} c_k(m)\alpha_s^k \ln^k z_n\right]. \tag{8.2.21}$$

Let us consider the coefficients $b_m\left(\omega^2, \omega_n^2\right)$ from equation (8.2.18) in more detail. Noting that

$$\left(\omega^2 + \omega_n^2\right) = M\left(\bar{\omega}^2, \omega_n^2\right) - \frac{3}{4}\Delta_R^2, \tag{8.2.22}$$

these coefficients can be presented as follows:

$$\begin{aligned} b_m\left(\omega^2, \omega_n^2\right) &= -\left(-\frac{3}{4}\right)^m \frac{1}{m} \frac{\left(\omega^2 + \omega_n^2\right)^m}{M^m\left(\bar{\omega}^2, \omega_n^2\right)} \\ &= -\left(-\frac{3}{4}\right)^m \frac{1}{m} \sum_{p=0}^{m} \binom{m}{p} M^{p-m}\left(\bar{\omega}^2, \omega_n^2\right)\left(-\frac{3}{4}\Delta_R^2\right)^{m-p} \\ &= -\left(-\frac{3}{4}\right)^m \frac{1}{m}\left[1 + \sum_{p=0}^{m-1} \binom{m}{p} M^{p-m}\left(\bar{\omega}^2, \omega_n^2\right)\left(-\frac{3}{4}\Delta_R^2\right)^{m-p}\right], \end{aligned} \tag{8.2.23}$$

with the usual definition of

$$\binom{m}{p} = \frac{m(m-1)...(m-p+1)}{p!}, \qquad \binom{m}{0} = \binom{m}{m} = 1 \tag{8.2.24}$$

are the binomials coefficients. Substituting the expression (8.2.23) into equation (8.2.21), it becomes

$$P'_{PT}(T) = \tilde{P}_{PT}(T) + P'_{PT}\left(\Delta_R^2; T\right), \tag{8.2.25}$$

where the two terms on the right hand side are

$$\begin{aligned} \tilde{P}_{PT}(T) = &- \frac{8}{\pi^2} \int_{\Lambda_{YM}}^{\infty} d\omega\, \omega^2\, T \\ &\times \sum_{n=-\infty}^{+\infty} \left[\sum_{m=2}^{\infty} \left(-\frac{3}{4}\right)^m \frac{\alpha_s^m}{m} \sum_{k=0}^{\infty} c_k(m) \alpha_s^k \ln^k z_n\right], \end{aligned} \tag{8.2.26}$$

and

$$\begin{aligned} P'_{PT}(\Delta_R^2; T) = &- \frac{8}{\pi^2} \int_{\Lambda_{YM}}^{\infty} d\omega\, \omega^2\, T \\ &\times \sum_{n=-\infty}^{+\infty} \left[\sum_{m=2}^{\infty} \left(-\frac{3}{4}\right)^m \frac{\alpha_s^m}{m} P_m^{(n)}(\Delta_R^2) \sum_{k=0}^{\infty} c_k(m) \alpha_s^k \ln^k z_n\right] \end{aligned} \tag{8.2.27}$$

with the parameters within the integral,

$$P_m^{(n)}\left(\Delta_R^2\right) = \sum_{p=0}^{m-1} \binom{m}{p} M^{p-m}\left(\bar{\omega}^2, \omega_n^2\right)\left(-\frac{3}{4}\Delta_R^2\right)^{m-p}. \tag{8.2.28}$$

It is convenient to present the integral (8.2.26) in the following way

$$\tilde{P}_{PT}(T) = -\frac{9}{2\pi^2}\alpha_s^2 \int_{\Lambda_{YM}}^{\infty} d\omega\, \omega^2\, T \times \sum_{n=-\infty}^{+\infty}\left[\sum_{m=0}^{\infty}\left(-\frac{3}{4}\right)^m \frac{\alpha_s^m}{m+2}\sum_{k=0}^{\infty} c_k(m+2)\alpha_s^k \ln^k z_n\right], \tag{8.2.29}$$

which shows explicitly that it is an α_s^2-order term.

It is also convenient to present the integral (8.2.27) in the same way,

$$P'_{PT}\left(\Delta_R^2; T\right) = -\frac{9}{2\pi^2}\alpha_s^2 \int_{\Lambda_{YM}}^{\infty} d\omega\, \omega^2\, T \times \sum_{n=-\infty}^{+\infty}\left[\sum_{m=0}^{\infty}\left(-\frac{3}{4}\right)^m \frac{\alpha_s^m}{m+2} P_{m+2}^{(n)}\left(\Delta_R^2\right)\sum_{k=0}^{\infty} c_k(m+2)\alpha_s^k \ln^k z_n\right] \tag{8.2.30}$$

with the parameter within the integral,

$$\begin{aligned} P_{m+2}^{(n)}\left(\Delta_R^2\right) &= \sum_{p=0}^{m+1}\binom{m+2}{p} M^{p-m-2}(\bar{\omega}^2, \omega_n^2)\left(-\frac{3}{4}\Delta_R^2\right)^{m+2-p} \\ &= \left(\frac{3}{4}\Delta_R^2\right)^2 M^{-2}(\bar{\omega}^2, \omega_n^2)\sum_{p=0}^{m+1}\binom{m+2}{p} M^{p-m}(\bar{\omega}^2, \omega_n^2)\left(-\frac{3}{4}\Delta_R^2\right)^{m-p} \\ &= \left(\frac{3}{4}\Delta_R^2\right)^2 P'^{(n)}_{m+2}(\Delta_R^2). \end{aligned} \tag{8.2.31}$$

Then the previous integral (8.2.30) becomes,

$$P'_{PT}\left(\Delta_R^2; T\right) = -\left(\frac{9\alpha_s\Delta_R^2}{\sqrt{24}\pi}\right)^2 \int_{\Lambda_{YM}}^{\infty} d\omega\, \omega^2\, T \times \sum_{n=-\infty}^{+\infty}\left[\sum_{m=0}^{\infty}\frac{(-3/4)^m\alpha_s^m}{m+2} P'^{(n)}_{m+2}\left(\Delta_R^2\right)\sum_{k=0}^{\infty} c_k(m+2)\alpha_s^k \ln^k z_n\right], \tag{8.2.32}$$

and in the formal $\Delta_R^2 = 0$ limit it is zero as is the whole expansion (8.2.13). The integral (8.2.32) is really of the order of α_s^2. However, this order is numerically much smaller than the corresponding order term in the expansion (8.2.13). This will be true for any corresponding orders in the expansions (8.2.13) and (8.2.32) because of the initial condition $x\left(\omega^2, \omega_n^2\right) \ll 1$ in equation (8.2.10). Obviously, the structures of all expansions (8.2.13), (8.2.29) and (8.2.32) differ from each other.

8.3 Convergence of the perturbation theory series

The convergence of the power series (8.2.10) over m is obvious, since it comes from the expansion of the logarithm (8.2.6). In turn, this means that the series in integer powers of α_s (8.2.29) and (8.2.32), due to the relation (8.2.25), are also convergent. At the same time, the convergence of the series (8.2.13) over k is not so obvious and requires some additional investigation. The series (8.2.13) can be formally considered as a power series over small α_s with the coefficients $P_k\left(\Delta^2; T\right)$. The QCD fine-structure coupling constant correctly calculated at any scale is indeed always small. Let us remind that in this book we use its value calculated at the Z-boson mass, namely $\alpha_s = \alpha_s(m_Z) = 0.1184$ based on [167]. For the convergence of the power series (8.2.13) it is necessary to show that its radius of convergence, r is bigger than any possible value of α_s. This r of the power series can be calculated in accordance with the Cauchy – Hadamard theorem as follows:

$$r^{-1} = \lim_{k\to\infty}\left|\frac{P_{k+1}\left(\Delta^2; T\right)}{P_k\left(\Delta^2; T\right)}\right| \tag{8.3.1}$$

if this limit exists. Substituting the corresponding expressions from equation (8.2.14), one obtains

$$r^{-1} = \lim_{k\to\infty}\left|\frac{\int\limits_{\Lambda_{YM}}^{\infty} d\omega\, \omega^2 \sum\limits_{n=-\infty}^{+\infty}\left[\frac{1}{M(\bar{\omega}^2,\omega_n^2)}\ln^k z_n\right]}{\int\limits_{\Lambda_{YM}}^{\infty} d\omega\, \omega^2 \sum\limits_{n=-\infty}^{+\infty}\left[\frac{1}{M(\bar{\omega}^2,\omega_n^2)}\ln^{k-1} z_n\right]}\right| \longrightarrow 1. \tag{8.3.2}$$

Thus the series (8.2.13) converges absolutely for $\alpha_s < r = 1$ and converges uniformly on every compact subset of $\{\alpha_s : \alpha_s < r\}$. Roughly speaking this means that any further calculated term in integer powers of α_s will be smaller than the previous one — at least by one order of magnitude. Let us also note that now there is no need in the restriction $|\alpha_s \ln z_n| < 1$; only $\alpha_s < 1$ is required, which always holds.

Summing up all the perturbative contributions, the perturbative pressure (8.2.1) finally becomes

$$\begin{aligned} P_{PT}(T) &= P_{PT}\left(\Delta_R^2;T\right) + P'_{PT}(T) \\ &= \left[P_{PT}\left(\Delta_R^2;T\right) + P'_{PT}\left(\Delta_R^2;T\right)\right] + \tilde{P}_{PT}(T). \end{aligned} \tag{8.3.3}$$

Due to this decomposition a few things should be made perfectly clear. All three integrals (7.3.6), (7.3.7), and (7.3.8) contributing to the gluon pressure (7.3.5) are the corresponding parts of the initial skeleton loop integral derived in the previous chapter and in Ref. [215] as well. The first two integrals depend on the mass gap Δ_R^2 only, so they are intrinsically non-perturbative, indeed. We call the integral (7.3.8) or, equivalently, (8.2.25) the perturbative one because it explicitly depends on α_s, but this is only a convention. Its two terms $P_{PT}\left(\Delta_R^2;T\right)$ given in equation (8.2.13) and $P'_{PT}\left(\Delta_R^2;T\right)$ by equation (8.2.32) depend on the mass gap Δ_R^2 as well — they vanish in the formal $\Delta_R^2 = 0$ limit. Its third term $\tilde{P}_{PT}(T)$ in equation (8.2.29) begins with the α_s^2-order correction, and does not depend on the mass gap at all. However, it is an infinite sum of perturbative contributions, thus, in fact this is the corresponding part of the skeleton loop term (8.2.25). The terms $P_{PT}\left(\Delta_R^2;T\right)$ and $P'_{PT}\left(\Delta_R^2;T\right)$, being by themselves the two other corresponding parts of the skeleton loop term (8.2.25), can be considered as the α_s- and α_s^2-order corrections to the skeleton loop term $P_{YM}(T)$ (7.3.7), since they depend on the massive gluonic excitation $\bar{\omega}$. Let us emphasize that there cannot be any perturbative corrections to the bag constant $B_{YM}(T)$, since its skeleton part (7.3.6) has been defined from the very beginning by the subtraction of all the types of the perturbative contributions. All three expansions (8.2.13), (8.2.29) and (8.2.32) analytically depend on α_s, and they can be calculated termwise in integer powers of α_s. The convergence of the series in integer powers of a small α_s derived here has been confirmed. It guarantees that any further perturbative contribution in powers of α_s will be numerically smaller than the previous one. It is worth emphasizing that in any case none of the perturbative contributions, and hence none of their sum, can be numerically bigger than the Stefan–Boltzmann term, which describes the thermodynamic structure of the gluon matter at high temperatures, as plotted on Figs. 8.4.1 and 8.7.6 below.

A few general remarks are in order. In the initial QCD thermal perturbation theory the dependence on α_s is non-analytical, i.e., the expansion contains its fractional powers, $\alpha_s^3 \ln \alpha_s$, etc. — see, for example Refs. [184, 191, 193, 213, 216]. This leads to the divergent series in the

QCD thermal perturbation theory, though each term up to the $\alpha_s^3 \ln \alpha_s$-order has been calculated correctly. In Ref. [215] it has briefly been explained why this effect occurs there. This problem does not appear in our formalism, which is non-perturbative from the very beginning as underlined above. Apparently, the reason is that it allows to calculate the α_s-dependent corrections, which by themselves are the corresponding skeleton parts of the skeleton loop integral (7.3.8), to the non-perturbative term (7.3.7) which is the skeleton part of the initial skeleton loop integral (7.3.5). In other words, the re-summation of an infinite number of the corresponding contributions has been already done for all of them.

So our approach allows us to develop the series in integer powers of α_s for the non-perturbative quantities, i.e. the coefficients of these infinite series are the non-perturbative quantities by themselves. Let us emphasize that these conventionally called perturbation theory series are, in fact, the so-called cluster convergent expansions, depending on the coupling constant in a non-elementary way [11]. This is completely different from the QCD thermal perturbation theory, which in the best case can be only of the asymptotic-type expansion. It is dealing with the pure perturbative contributions from the very beginning, and the necessary re-summation of the hard thermal loops [184] leads finally to the non-analytical dependence on α_s. At the same time, the formalism developed here makes it possible to calculate the α_s-dependent contributions to the gluon pressure in terms of the convergent series in integer powers of a small α_s. This seems to resolve the above-mentioned long-term problem.

However, instead it may produce another problem, namely the question of double-counting in α_s. In the present chapter there is no double-counting in α_s, because all the convergent perturbative series (8.2.13), (8.2.29), and (8.2.32) are of different structure, as emphasized above. The series (8.2.29) and (8.2.32) begin with the α_s^2-orders, and therefore their numerical contributions are very small, since the series are convergent. The convergent series (8.2.13) begins with the α_s-order, which is the only one to be numerically calculated below. The double-counting problem will indeed arise when we include the Stefan – Boltzmann term into the full equation of state, since it should be approached in the asymptotic freedom way, which involves all powers of α_s. So the question of how far the pure perturbative corrections overlap with the non-perturbative ones cannot be answered here. But the problem is fixed, and we will address it and clarify the situation in the forthcoming Chapter.

8.4 The gluon pressure, $P_g(T)$

Summing up all contributions, the gluon pressure given by equation (7.3.5) thus finally becomes

$$\begin{aligned} P_g(T) &= P_{NP}(T) + P_{PT}(T) \\ &= P_{NP}(T) + [P_{PT}(\Delta_R^2;T) + P'_{PT}(\Delta_R^2;T)] + \tilde{P}_{PT}(T). \end{aligned} \quad (8.4.1)$$

In general, both expansions $P_{PT}\left(\Delta_R^2;T\right)$ and $P'_{PT}\left(\Delta_R^2;T\right)$ (8.2.13) (8.2.32) are to be considered as producing the corresponding perturbative corrections to the leading non-perturbative part $P_{NP}(T)$ of the gluon matter equation of state by equation (8.4.1). At the same time, the pure perturbative term $\tilde{P}_{PT}(T)$ (8.2.29) is to be considered as producing the perturbative corrections to the leading perturbative contribution which is nothing but the above-mentioned Stefan – Boltzmann term. However, due to the normalization condition of the free perturbative vacuum to zero, since normalization of $P_g(T)$ is to zero when the interaction is switched off. This is not *explicitly* present in equation (8.4.1). *So none of pure perturbative corrections has to be calculated unless the leading Stefan – Boltzmann term is restored to equation* (8.4.1) *in a self-consistent way.* In this connection, let us note that the Stefan – Boltzmann term or, equivalently, the pressure of a gas of massless free gluons (ideal gas) can be considered as the $\alpha_s^0 = 1$-order pure perturbative contribution to the full pressure. That is why we start the numerical calculation of the perturbative contributions to the gluon pressure (8.4.1) from its first non-trivial order, namely the α_s-order in the expansion (8.2.13). As emphasized above, it is the α_s-order contribution to $P_{YM}(T)$ in the non-perturbative term, which is already present in the gluon matter equation of state (8.4.1). Then it looks like

$$P_g(T) = P_{NP}(T) + P^s_{PT}(T) + \mathcal{O}\left(\alpha_s^2\right), \quad (8.4.2)$$

where $P^s_{PT}(T) = \alpha_s P_1\left(\Delta_R^2;T\right)$ and for $P_1\left(\Delta_R^2;T\right)$ see equation (8.2.16). Omitting the terms of the $\mathcal{O}\left(\alpha_s^2\right)$-order, for convenience, it is instructive to explicitly gather all our results from the relations (7.5.7)–(7.5.9) and (8.2.15) for the gluon pressure (8.4.2) once more as follows:

$$P_g(T) = \frac{6}{\pi^2}\Delta_R^2 P_1(T) + \frac{16}{\pi^2}T\left[P_2(T) + P_3(T) - P_4(T)\right] + P^s_{PT}(T), \quad (8.4.3)$$

where the integral $P^s_{PT}(T)$ is given in the form,

$$P^s_{PT}(T) = \alpha_s \cdot \frac{9}{2\pi^2}\Delta_R^2 \int\limits_{\Lambda_{YM}}^{\infty} d\omega\, \omega^2\, \frac{1}{\bar{\omega}}\frac{1}{e^{\beta\bar{\omega}} - 1}, \quad (8.4.4)$$

while all other integrals $P_n(T)$, $n = 1, 2, 3, 4$ are given in equations (7.5.12) and (7.5.13). This form is convenient for the numerical calculations. Let us note that when the interaction is formally switched off, i.e., letting $\alpha_s = \Delta_R^2 = 0$, the composition $[P_2(T) + P_3(T) - P_4(T)]$ becomes identically zero (see equations (7.5.13)) and thus $P_g(T)$ itself. The gluon pressure (8.4.3), the non-perturbative pressure $P_{NP}(T)$ and its first perturbative contribution of the α_s order $P_{PT}^s(T)$ (8.4.4) are also shown in Fig. 8.4.1. However, it is worth emphasizing that, in fact, the term $P_{PT}^s(T)$ is non-perturbative, depending on the mass gap Δ_R^2, which is only suppressed by the α_s order.

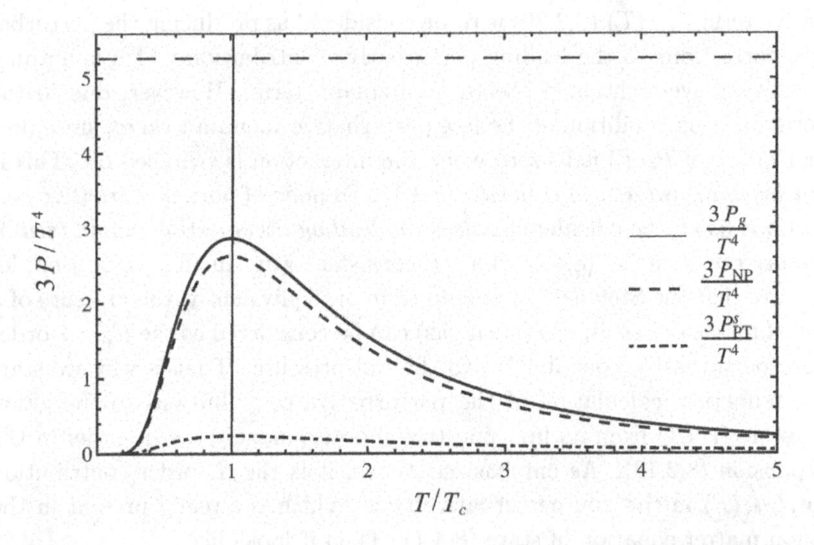

Fig. 8.4.1 The gluon pressure (8.4.3), the non-perturbative pressure (7.5.11) and the α_s-dependent perturbative pressure (8.4.4), all scaled by $T^4/3$, are shown as functions of T/T_c. Effectively, all curves have maxima at $T_c = 266.5$ MeV (*vertical solid line*). The horizontal dashed line is the Stefan–Boltzmann constant $3P_{SB}(T)/T^4 = (24/45)\pi^2 \approx 5.2634$.

8.5 Low-temperature expansion

Let us begin with noting in advance that all exactly calculated integrals discussed below can be found in Refs. [214, 217]. This is also true for the asymptotics in the low- and high-temperature limits of those integrals which cannot be analytically derived. In order to evaluate a low-temperature expansion for the gluon pressure

$$P_g(T) = P_{NP}(T) + P_{PT}^s(T), \tag{8.5.1}$$

it is convenient to present the non-perturbative pressure as in equation (7.5.7), namely

$$P_{NP}(T) = \frac{6}{\pi^2}\Delta_R^2 P_1(T) + \frac{16}{\pi^2} T M(T), \tag{8.5.2}$$

where the $P_1(T)$ from equation (7.5.12) is

$$P_1(T) = \int_{\omega_{eff}}^{\infty} d\omega \frac{\omega}{e^{\beta\omega} - 1}, \tag{8.5.3}$$

and

$$M(T) = P_2(T) + P_3(T) - P_4(T), \tag{8.5.4}$$

with

$$P_2(T) = \int_{\omega_{eff}}^{\infty} d\omega\ \omega^2 \ln\left[1 - e^{-\beta\omega}\right],$$

$$P_3(T) = \int_{0}^{\omega_{eff}} d\omega\ \omega^2 \ln\left[1 - e^{-\beta\omega'}\right],$$

$$P_4(T) = \int_{0}^{\infty} d\omega\ \omega^2 \ln\left[1 - e^{-\beta\bar{\omega}}\right]. \tag{8.5.5}$$

In all the above-displayed integrals the variable $y = e^{-\beta\omega}$ is always small, and hence $y^{-1} = e^{\beta\omega}$ is always large, in the low-temperature limit $T \to 0$ ($\beta = T^{-1} \to \infty$). This is true for the exponents $e^{-\beta\omega'}$ and $e^{-\beta\bar{\omega}}$ as well. So to the leading order term, $P_1(T)$ becomes

$$P_1(T) \sim \int_{\omega_{eff}}^{\infty} d\omega\ \omega e^{-\beta\omega}. \tag{8.5.6}$$

The almost trivial integration of this yields

$$P_1(T) \sim \left[T^2 + \omega_{eff} T\right] e^{-\frac{\omega_{eff}}{T}}, \quad \text{with } T \to 0. \tag{8.5.7}$$

The integral $P_2(T)$ can be considered in the same way. Up to the leading order it becomes

$$P_2(T) = \int_{\omega_{eff}}^{\infty} d\omega\ \omega^2 \ln\left[1 - e^{-\beta\omega}\right] \sim - \int_{\omega_{eff}}^{\infty} d\omega\ \omega^2\ e^{-\beta\omega}, \text{ with } \beta \to \infty, \tag{8.5.8}$$

and integrating it, one obtains

$$P_2(T) \sim -\left[2T^3 + 2\omega_{eff}T^2 + \omega_{eff}^2 T\right] e^{-\frac{\omega_{eff}}{T}}, \text{ with } T \to 0. \tag{8.5.9}$$

The integral $P_3(T)$ up to the leading order looks like

$$P_3(T) = \int_0^{\omega_{eff}} d\omega\, \omega^2 \ln\left[1 - e^{-\beta\omega'}\right] \sim -\int_0^{\omega_{eff}} d\omega\, \omega^2 e^{-\beta\omega'}, \text{ with } \beta \to \infty, \tag{8.5.10}$$

and replacing the variable ω by the variable ω' in accordance with the relation (7.5.3), this integral becomes

$$P_3(T) \sim -\int_a^{\omega'_{eff}} d\omega'\, \omega' \sqrt{\omega'^2 - a^2}\, e^{-\beta\omega'}, \text{ with } \beta \to \infty, \tag{8.5.11}$$

where

$$\omega'_{eff} = \sqrt{\omega_{eff}^2 + a^2} \quad \text{with} \quad a = \sqrt{3}\Delta_R. \tag{8.5.12}$$

Unfortunately, this asymptotical expression cannot be directly evaluated. However, noting that the variable $x = a^2/\omega'^2 \leq 1$, we can formally expand

$$\sqrt{\omega'^2 - a^2} = \omega'(1-x)^{1/2} = \omega'\left[1 - \frac{1}{2}\frac{a^2}{\omega'^2} + \sum_{k=2}^{\infty}\binom{1/2}{k}(-x)^k\right]. \tag{8.5.13}$$

Then from the integral (8.5.11) one obtains

$$P_3(T) \sim -\int_{\sqrt{3}\Delta_R}^{\omega'_{eff}} d\omega'\, \omega'^2\, e^{-\beta\omega'} + \frac{3}{2}\Delta_R^2 \int_{\sqrt{3}\Delta_R}^{\omega'_{eff}} d\omega'\, e^{-\beta\omega'} + P_3^{(k)}(T), \quad \beta \to \infty, \tag{8.5.14}$$

where the last term is given by

$$P_3^{(k)}(T) = -\int_{\sqrt{3}\Delta_R}^{\omega'_{eff}} d\omega'\, \omega'^2\, e^{-\beta\omega'} \sum_{k=2}^{\infty}\binom{1/2}{k}(-x)^k. \tag{8.5.15}$$

Let us consider the last integral in more detail. Since the series over k are convergent in the interval of integration and the functions depending on k are integrable in this interval, these series may be integrated termwise [214], that is,

$$P_3^{(k)}(T) = -\sum_{k=2}^{\infty}\binom{1/2}{k}(-a^2)^k \int_{\sqrt{3}\Delta_R}^{\omega'_{eff}} d\omega'\, \frac{e^{-\beta\omega'}}{(\omega')^{2k-2}}. \tag{8.5.16}$$

Integrating it, one obtains

$$P_3^{(k)}(T) = -\sum_{k=2}^{\infty} \binom{1/2}{k} (-a^2)^k \left[N_3^{(k)}(T, \omega') \right]_{\sqrt{3}\Delta_R}^{\omega'_{eff}} \quad (8.5.17)$$

and $\left[N_3^{(k)}(T, \omega') \right]_{\sqrt{3}\Delta_R}^{\omega'_{eff}}$ denotes the result of the integration over ω' in equation (8.5.16) on the interval $\left[\sqrt{3}\Delta_R, \omega'_{eff} \right]$, while the function $N_3^{(k)}(T, \omega')$ itself is

$$N_3^{(k)}(T, \omega') = -e^{-\beta\omega'} \times \sum_{m=1}^{2k-3} \frac{(-\beta)^{m-1}(\omega')^{m+2-2k}}{(2k-3)(2k-4)...(2k-2-m)} + \frac{(-\beta)^{2k-3}}{(2k-3)!} \mathrm{Ei}(-\beta\omega'). \quad (8.5.18)$$

The series for the exponential integral function $\mathrm{Ei}(-\beta\omega')$ is

$$\mathrm{Ei}(-\beta\omega') = e^{-\beta\omega'} \sum_{l=1}^{n} (-1)^l \frac{(l-1)!}{(\beta\omega')^l}, \quad \text{where } n \geq 2k-3 \text{ and } \beta \to \infty. \quad (8.5.19)$$

If one chooses $n = 2k - 3$ in the previous equation and neglecting the relative error of the approximation (8.5.19), it is easy to show that both terms in equation (8.5.18) for $N_3^{(k)}(T, \omega')$ cancel each other termwise for any $k \geq 2$, and thus

$$N_3^{(k)}(T, \omega') = 0, \quad k = 2, 3, 4, ... \quad (8.5.20)$$

or, equivalently,

$$P_3^{(k)}(T) = 0, \quad k = 2, 3, 4, ... \quad (8.5.21)$$

Going back to equation (8.5.14) and easily integrating the first two terms, and taking into account the previous result, one comes to the following expansion

$$\begin{aligned} P_3(T) \sim & \left[2T^3 + 2\omega'_{eff}T^2 + \omega'^2_{eff}T \right] e^{-\frac{\omega'_{eff}}{T}} \\ & - \left[2T^3 + 2\sqrt{3}\Delta_R T^2 + 3\Delta_R^2 T \right] e^{-\frac{\sqrt{3}\Delta_R}{T}} \\ & - \frac{3}{2}\Delta_R^2 T e^{-\frac{\omega'_{eff}}{T}} + \frac{3}{2}\Delta_R^2 T e^{-\frac{\sqrt{3}\Delta_R}{T}}, \quad \text{where } T \to 0. \end{aligned} \quad (8.5.22)$$

The integral $P_4(T)$ to the leading order looks like

$$P_4(T) = \int_0^\infty d\omega\, \omega^2 \ln\left[1 - e^{-\beta\bar{\omega}}\right] \sim - \int_0^\infty d\omega\, \omega^2 e^{-\beta\bar{\omega}}, \text{ with } \beta \to \infty, \tag{8.5.23}$$

and replacing the variable ω by the variable $\bar{\omega}$ in accordance with the relation (7.5.8), this integral becomes

$$P_4(T) \sim - \int_{(a/2)}^\infty d\bar{\omega}\, \bar{\omega}\sqrt{\bar{\omega}^2 - (a/2)^2}\, e^{-\beta\bar{\omega}}, \quad \text{with } \beta \to \infty. \tag{8.5.24}$$

Noting again that the variable $z = a^2/4\bar{\omega}^2 \leq 1$, we can formally expand

$$\sqrt{\bar{\omega}^2 - (a/2)^2} = \bar{\omega}\left[1 - \frac{a^2}{8\bar{\omega}^2} + \sum_{k=2}^\infty \binom{1/2}{k}(-z)^k\right]. \tag{8.5.25}$$

Then from the integral (8.5.24) one obtains

$$P_4(T) \sim - \int_{\sqrt{3}\Delta_R/2}^\infty d\bar{\omega}\, \bar{\omega}^2\, e^{-\beta\bar{\omega}} + \frac{3}{8}\Delta_R^2 \int_{\sqrt{3}\Delta_R/2}^\infty d\bar{\omega}\, e^{-\beta\bar{\omega}} + P_4^{(k)}(T), \tag{8.5.26}$$

where next to the $\beta \to \infty$ assumption, due to the same formalism which has been used previously in order to get the result (8.5.21), one can conclude that $P_4^{(k)}(T) = 0$ as well. Easily integrating the first two terms, one comes to the following expansion, taking into account $T \to 0$,

$$P_4(T) \sim - \left[2T^3 + \sqrt{3}\Delta_R T^2 + \frac{3}{4}\Delta_R^2 T\right] e^{-\frac{\sqrt{3}\Delta_R}{2T}} + \frac{3}{8}\Delta_R^2 T e^{-\frac{\sqrt{3}\Delta_R}{2T}}. \tag{8.5.27}$$

Substituting all these expansions into the equation (8.5.2), one obtains

$$\begin{aligned}
P_{NP}(T) \sim\ & \frac{6}{\pi^2}\Delta_R^2\left[T^2 + \omega_{eff}T\right] e^{-\frac{\omega_{eff}}{T}} \\
& - \frac{16}{\pi^2}T\left[2T^3 + 2\omega_{eff}T^2 + \omega_{eff}^2 T\right] e^{-\frac{\omega_{eff}}{T}} \\
& + \frac{16}{\pi^2}T\left[2T^3 + 2\omega'_{eff}T^2 + \omega'^2_{eff}T\right] e^{-\frac{\omega'_{eff}}{T}} \\
& - \frac{16}{\pi^2}T\left[2T^3 + 2\sqrt{3}\Delta_R T^2 + 3\Delta_R^2 T\right] e^{-\frac{\sqrt{3}\Delta_R}{T}} \\
& - \frac{24}{\pi^2}T^2\Delta_R^2\left[e^{-\frac{\omega'_{eff}}{T}} - e^{-\frac{\sqrt{3}\Delta_R}{T}}\right] \\
& + \frac{16}{\pi^2}T\left[2T^3 + \sqrt{3}\Delta_R T^2 + \frac{3}{8}\Delta_R^2 T\right] e^{-\frac{\sqrt{3}\Delta_R}{2T}}, \quad \text{with } T \to 0.
\end{aligned} \tag{8.5.28}$$

Evidently, this is nothing but a low-temperature expansion for the pressure $P_{NP}(T)$. Let us note that it contains a non-analytical dependence on the mass gap squared in terms $\sim (\Delta_R^2)^{1/2}T^3 \sim \Delta_R T^3$, but the mass gap is not an expansion parameter like α_s.

It is instructive to re-write this expansion as follows:

$$\begin{aligned}
P_{NP}(T) &\sim F_{NP}(\Delta_R, T) \\
&+ \frac{180}{\pi^4}\left[-e^{-\frac{\omega_{eff}}{T}} + e^{-\frac{\omega'_{eff}}{T}} - e^{-\frac{\sqrt{3}\Delta_R}{T}} + e^{-\frac{\sqrt{3}\Delta_R}{2T}}\right] P_{SB}(T) \\
&- \frac{16}{\pi^2}T^2\left[2\omega_{eff}T + \omega_{eff}^2\right]e^{-\frac{\omega_{eff}}{T}} \\
&+ \frac{16}{\pi^2}T^2\left[2\omega'_{eff}T + \omega'^2_{eff}\right]e^{-\frac{\omega'_{eff}}{T}}, \quad \text{with } T \to 0, \qquad (8.5.29)
\end{aligned}$$

since for the Yang – Mills fields based on Refs. [189, 184] one can take

$$P_{SB}(T) = \frac{8}{45}\pi^2 T^4. \qquad (8.5.30)$$

$F_{NP}(\Delta_R, T)$ denotes the sum of all the terms which directly depend on the mass gap, so that $F_{NP}(T, \Delta_R = 0) = 0$. Thus the propagation of massless gluons below T_c can be described by the Stefan – Boltzmann-type term. However, it is exponentially suppressed in the $T \to 0$ limit, as it should be. At $T \sim T_c$ its contribution can be numerically comparable with other contributions in equation (8.5.28). It is no surprise that the massless gluons may be present in the gluon matter at any temperature. Moreover, let us note in advance that the propagation of the massless free gluons below T_c (if any) cannot be described by the Stefan – Boltzmann term itself. It should also be exponentially suppressed in the same way as it is shown in equation (8.5.29). But it has to survive in the perturbative $\Delta_R^2 = 0$ limit, while the contribution (8.5.29) vanishes in this limit — when $\omega'_{eff} = \omega_{eff}$ in this case, see equation (8.5.12). The Stefan – Boltzmann term can describe the propagation of the massless free gluons only in the high temperatures limit above T_c.

Let us now consider equation (8.4.4), which in the $T \to 0$ limit to the leading order becomes

$$P_{PT}^s(T) = \frac{9\alpha_s}{2\pi^2}\Delta_R^2 \int_{\Lambda_{YM}}^{\infty} d\omega\, \omega^2\, \frac{1}{\bar{\omega}}\frac{1}{e^{\beta\bar{\omega}} - 1} \sim \frac{9\alpha_s}{2\pi^2}\Delta_R^2 \int_{\Lambda_{YM}}^{\infty} d\omega\, \omega^2\, \frac{1}{\bar{\omega}}e^{-\beta\bar{\omega}}, \qquad (8.5.31)$$

and $\bar{\omega}$ is given by the relation (7.5.8). Replacing the variable ω by the variable $\bar{\omega}$, as in equation (8.5.23), one obtains

$$P_{PT}^{s}(T) \sim \frac{9\alpha_s}{2\pi^2}\Delta_R^2 \int_{\tilde{\omega}_{eff}}^{\infty} d\bar{\omega}\, \sqrt{\bar{\omega}^2 - (a/2)^2}\, e^{-\beta\bar{\omega}}, \quad \beta \to \infty, \tag{8.5.32}$$

where $\tilde{\omega}_{eff} = \sqrt{\Lambda_{YM}^2 + (a/2)^2}$, and for a see equation (8.5.12). Noting that the variable $z = a^2/4\bar{\omega}^2 < 1$ in this case, we can use the expansion (8.5.25) at $\beta \to \infty$ in order to obtain

$$P_{PT}^{s}(T) \sim \frac{9\alpha_s}{2\pi^2}\Delta_R^2 \left[\int_{\tilde{\omega}_{eff}}^{\infty} d\bar{\omega}\, \bar{\omega}\, e^{-\beta\bar{\omega}} - \frac{3}{8}\Delta_R^2 \int_{\tilde{\omega}_{eff}}^{\infty} d\bar{\omega}\, \frac{e^{-\beta\bar{\omega}}}{\bar{\omega}} + P_s^{(k)}(T) \right]. \tag{8.5.33}$$

Due to the same formalism which has been used previously in order to get the result (8.5.21), one can conclude that $P_s^{(k)}(T) = 0$ as well. Easily integrating the first two terms, one comes to the following expansion

$$P_{PT}^{s}(T) \sim \frac{9\alpha_s}{2\pi^2}\Delta_R^2 \left[\left(T^2 + T\tilde{\omega}_{eff}\right) e^{-\frac{\tilde{\omega}_{eff}}{T}} + \frac{3}{8}\Delta_R^2 \mathrm{Ei}\left(-\frac{\tilde{\omega}_{eff}}{T}\right)\right], \quad T \to 0. \tag{8.5.34}$$

Here and below the corresponding exponential integral functions are defined by equation (8.5.19). Summing up the expansions (8.5.28) and (8.5.34), one obtains a low-temperature expansion in the $T \to 0$ limit, for the gluon pressure (8.5.1) as follows:

$$\begin{aligned}
P_g(T) \sim\ & \frac{6}{\pi^2}\Delta_R^2 \left[T^2 + \omega_{eff}T\right] e^{-\frac{\omega_{eff}}{T}} \\
& - \frac{16}{\pi^2} T \left[2T^3 + 2\omega_{eff}T^2 + \omega_{eff}^2 T\right] e^{-\frac{\omega_{eff}}{T}} \\
& + \frac{16}{\pi^2} T \left[2T^3 + 2\omega'_{eff}T^2 + \omega'^2_{eff} T\right] e^{-\frac{\omega'_{eff}}{T}} \\
& - \frac{16}{\pi^2} T \left[2T^3 + 2\sqrt{3}\Delta_R T^2 + 3\Delta_R^2 T\right] e^{-\frac{\sqrt{3}\Delta_R}{T}} \\
& - \frac{24}{\pi^2} T^2 \Delta_R^2 \left[e^{-\frac{\omega'_{eff}}{T}} - e^{-\frac{\sqrt{3}\Delta_R}{T}}\right] \\
& + \frac{16}{\pi^2} T \left[2T^3 + \sqrt{3}\Delta_R T^2 + \frac{3}{8}\Delta_R^2 T\right] e^{-\frac{\sqrt{3}\Delta_R}{2T}} \\
& + \frac{9\alpha_s}{2\pi^2}\Delta_R^2 \left[\left(T^2 + T\tilde{\omega}_{eff}\right) e^{-\frac{\tilde{\omega}_{eff}}{T}} + \frac{3}{8}\Delta_R^2 \mathrm{Ei}\left(-\frac{\tilde{\omega}_{eff}}{T}\right)\right].
\end{aligned} \tag{8.5.35}$$

Let us note that the low-temperature expansion (8.5.35) depends mainly on the effective massive 'excitation' $\omega'_{eff} = \sqrt{\omega^2_{eff} + a^2}$, and does not depend on the effective massive 'excitation' $\bar{\omega}_{eff} = \sqrt{\omega^2_{eff} + (a/2)^2}$ at all, while the dependence on the effective massive excitation $\tilde{\omega}_{eff} = \sqrt{\Lambda^2_{YM} + (a/2)^2}$ is suppressed by α_s.

It is instructive to use in the exponents of the previous expansion the following obvious relations: $\omega_{eff} = \nu_1 T_c$, $\omega'_{eff} = \nu_2 T_c$, $\sqrt{3}\Delta_R = \nu_3 T_c$, $\nu_4 = (1/2)\nu_3$, $\tilde{\omega}_{eff} = \nu_5 T_c$, since all numerical values of these parameters are known. Then the previous expansion looks like in the same $T \to 0$ limit,

$$\begin{aligned}
P_g(T) \sim\ & \frac{6}{\pi^2}\Delta_R^2\left[T^2 + \omega_{eff}T\right]e^{-\nu_1\frac{T_c}{T}} \\
& - \frac{16}{\pi^2}T\left[2T^3 + 2\omega_{eff}T^2 + \omega^2_{eff}T\right]e^{-\nu_1\frac{T_c}{T}} \\
& + \frac{16}{\pi^2}T\left[2T^3 + 2\omega'_{eff}T^2 + \omega'^2_{eff}T\right]e^{-\nu_2\frac{T_c}{T}} \\
& - \frac{16}{\pi^2}T\left[2T^3 + 2\sqrt{3}\Delta_R T^2 + 3\Delta_R^2 T\right]e^{-\nu_3\frac{T_c}{T}} \\
& - \frac{24}{\pi^2}T^2\Delta_R^2\left[e^{-\nu_2\frac{T_c}{T}} - e^{-\nu_3\frac{T_c}{T}}\right] \\
& + \frac{16}{\pi^2}T\left[2T^3 + \sqrt{3}\Delta_R T^2 + \frac{3}{8}\Delta_R^2 T\right]e^{-\nu_4\frac{T_c}{T}} \\
& + \frac{9\alpha_s}{2\pi^2}\Delta_R^2\left[\left(T^2 + T\tilde{\omega}_{eff}\right)e^{-\nu_5\frac{T_c}{T}} + \frac{3}{8}\Delta_R^2\mathrm{Ei}\left(-\nu_5\frac{T_c}{T}\right)\right].
\end{aligned} \tag{8.5.36}$$

The expansion (8.5.36) clearly shows that the exponential suppression of any pressure at low temperature below T_c is determined by the corresponding asymptotic of the gluon mean number based on Ref. [184], namely

$$N_g = \frac{1}{e^{\beta\omega} - 1} \sim e^{-\nu\frac{T_c}{T}}, \quad \text{at} \quad T_c > T = \beta^{-1} \to 0, \tag{8.5.37}$$

replacing ω by νT_c in each different case, as it is seen in the previous low-temperature expansion for the gluon pressure. For the scaled gluon pressure $3P_g(T)/T^4$ the expansion (8.5.36) is especially useful, since it depends on the dimensionless variable (T/T_c) only, and it is shown in Fig. 8.4.1 below T_c. The expansion (8.5.36) clearly shows that near to T_c the number of effective gluonic degrees of freedom and their magnitudes will be drastically increased, though the gluon pressure initially contains only five different massive and massless gluonic excitations.

8.6 High-temperature expansion

In order to evaluate a high-temperature expansion for the gluon pressure (8.5.1), it is convenient to present the non-perutbative pressure (8.5.2) as follows:

$$P_{NP}(T) = \Delta_R^2 T^2 - \frac{6}{\pi^2}\Delta_R^2 P_1'(T) + \frac{16}{\pi^2} T M(T), \tag{8.6.1}$$

since the middle term is

$$P_1(T) = \int_{\omega_{eff}}^{\infty} d\omega\ \omega\ N_g(\beta,\omega) = \int_0^{\infty} d\omega\ \omega N_g(\beta,\omega) - \int_0^{\omega_{eff}} d\omega\ \omega\ N_g(\beta,\omega)$$
$$= \frac{\pi^2}{6} T^2 - P_1'(T), \tag{8.6.2}$$

where one can use the following evaluation,

$$\int_0^{\infty} d\omega\ \omega\ N_g(\beta,\omega) = \int_0^{\infty} d\omega \frac{\omega}{e^{\beta\omega}-1} = \frac{\pi^2}{6} T^2, \tag{8.6.3}$$

and

$$P_1'(T) = \int_0^{\omega_{eff}} d\omega\ \omega\ N_g(\beta,\omega) = \int_0^{\omega_{eff}} d\omega \frac{\omega}{e^{\beta\omega}-1}, \tag{8.6.4}$$

and the composition $M(T)$ is already given by the relations (8.5.4) and (8.5.5).

In the high-temperature limit $\beta = T^{-1} \to 0$, the gluon mean number $N_g(\beta,\omega)$ in the integral (8.6.4) can be approximated by the corresponding series in powers of $(\beta\omega)$, since the variable ω is restricted,

$$N_g(\beta,\omega) = \frac{1}{e^{\beta\omega}-1} = (\beta\omega)^{-1}\left[1 - \frac{1}{2}(\beta\omega) + O(\beta^2)\right], \quad \beta \to 0. \tag{8.6.5}$$

Let us also note in advance that in what follows for our purpose it is sufficient to keep only the positive powers of T in the evaluation of the high-temperature expansion for the gluon pressure (8.5.1), and hence for each term in equation (8.5.1). Thus, for the asymptotic of the integral $P_1'(T)$ to the leading order in powers of T, one obtains

$$P_1'(T) = \int_0^{\omega_{eff}} d\omega \frac{\omega}{e^{\beta\omega}-1} \sim T\omega_{eff}, \quad T \to \infty. \tag{8.6.6}$$

In order to investigate the behavior of the composition $M(T)$ (8.5.4) at high temperature, it is convenient to decompose its integral $P_2(T)$, shown in equations (8.5.5), as follows:

$$P_2(T) = P_2^{(1)}(T) - P_2^{(2)}(T), \tag{8.6.7}$$

where the two terms are

$$P_2^{(1)}(T) = \int_0^\infty d\omega\, \omega^2 \ln\left[1 - e^{-\beta\omega}\right] = -\frac{\pi^4}{45}T^3 = -\frac{\pi^2}{8T}P_{SB}(T),$$

$$P_2^{(2)}(T) = \int_0^{\omega_{eff}} d\omega\, \omega^2 \ln\left[1 - e^{-\beta\omega}\right], \tag{8.6.8}$$

due to the relation (8.5.30). Let us note in advance that we will not need the high-temperature asymptotic of the integral $P_2^{(2)}(T)$.

The integral $P_3(T)$ to the leading order in powers of $\beta \to 0$ becomes

$$P_3(T) = \int_0^{\omega_{eff}} d\omega\, \omega^2 \ln\left[1 - e^{-\beta\omega'}\right] \sim \int_0^{\omega_{eff}} d\omega\, \omega^2 \ln \beta\omega', \tag{8.6.9}$$

in accordance with the expansion (8.6.5), since the variable ω is restricted, and hence the variable $\omega' = \sqrt{\omega^2 + a^2}$ as well, where $a = \sqrt{3}\Delta_R$. The last integral can be exactly calculated and the high-temperature expansion, $T \to \infty$ for $P_3(T)$ becomes

$$P_3(T) \sim \frac{1}{6}\omega_{eff}^3 \ln\left(\frac{\omega_{eff}'^2}{T^2}\right) - \frac{1}{9}\omega_{eff}^3 + \Delta_R^2\omega_{eff} - \sqrt{3}\Delta_R^3 \arctan\left(\frac{\omega_{eff}}{\sqrt{3}\Delta_R}\right), \tag{8.6.10}$$

where $\omega_{eff}'^2$ is given in equation (8.5.12).

It is convenient to decompose the integral $P_4(T)$ as the sum of two terms,

$$P_4(T) = P_4^{(1)}(T) + P_4^{(2)}(T), \tag{8.6.11}$$

where terms are

$$P_4^{(1)}(T) = \int_{\omega_{eff}}^\infty d\omega\, \omega^2 \ln\left[1 - e^{-\beta\bar{\omega}}\right],$$

$$P_4^{(2)}(T) = \int_0^{\omega_{eff}} d\omega\, \omega^2 \ln\left[1 - e^{-\beta\bar{\omega}}\right]. \tag{8.6.12}$$

Let us begin with $P_4^{(1)}(T)$, which can be re-written as follows:

$$P_4^{(1)}(T) = \int_{\omega_{eff}}^{\infty} d\omega \, \omega^2 \ln \left[1 - e^{-\beta\omega\sqrt{1+(a^2/4\omega^2)}} \right], \tag{8.6.13}$$

on account of the relation, namely $\bar{\omega} = \sqrt{\omega^2 + (3/4)\Delta_R^2} = \sqrt{\omega^2 + (a/2)^2}$. Since the variable ω is large, then $x = (a^2/4\omega^2) \ll 1$, and thus we can expand

$$\sqrt{(1+x)} = 1 + \frac{1}{2}x + \mathcal{O}(x^2), \quad \text{if} \quad x \to 0. \tag{8.6.14}$$

Then the integral (8.6.13) to the leading order in powers of small β becomes

$$P_4^{(1)}(T) \sim \int_{\omega_{eff}}^{\infty} d\omega \, \omega^2 \ln \left[1 - e^{-\beta\omega} e^{-(x\beta\omega/2)} \right], \tag{8.6.15}$$

where the argument of the exponent $(x\beta\omega/2) = (a^2/8\omega T) = z \ll 1$ in the $T, \omega \to \infty$ limit, so the integral (8.6.15) can be presented as follows:

$$P_4^{(1)}(T) \sim \int_{\omega_{eff}}^{\infty} d\omega \, \omega^2 \ln \left[1 - e^{-\beta\omega} \left(1 - z + \mathcal{O}(z^2) \right) \right], \text{ at } z \ll 1 \text{ and } \beta \to 0, \tag{8.6.16}$$

or, equivalently,

$$\begin{aligned} P_4^{(1)}(T) &\sim \int_{\omega_{eff}}^{\infty} d\omega \, \omega^2 \ln \left[\left(1 - e^{-\beta\omega} \right) \left(1 + \frac{z}{e^{\beta\omega} - 1} \right) \right] \\ &\sim P_2(T) + \int_{\omega_{eff}}^{\infty} d\omega \, \omega^2 \ln \left[1 + \frac{z}{e^{\beta\omega} - 1} \right] \\ &\sim P_2(T) + P_2'(T), \end{aligned} \tag{8.6.17}$$

as it follows from equations (8.5.5) or (8.6.7)–(8.6.8). The argument of the logarithm in the second integral is again always small $\left(z/e^{\beta\omega} - 1 \right) \ll 1$ in the T, $\omega \to \infty$ limit, and thus we can expand it and obtain in the leading order

$$P_2'(T) = \int_{\omega_{eff}}^{\infty} d\omega \, \omega^2 \ln \left[1 + \frac{z}{e^{\beta\omega} - 1} \right] \sim \frac{a^2}{8} \beta \int_{\omega_{eff}}^{\infty} d\omega \, \frac{\omega}{e^{\beta\omega} - 1}. \tag{8.6.18}$$

The last integral is nothing but $P_1(T)$ defined in equation (8.6.2), so that combining (8.6.2) and (8.6.6), one obtains

$$P_2'(T) \sim \frac{a^2}{8}\beta P_1(T) \sim \frac{\pi^2}{16}\Delta_R^2 T - \frac{3}{8}\Delta_R^2 \omega_{eff}, \quad \text{as} \quad T \to \infty, \tag{8.6.19}$$

and then the high-temperature expansion for $P_4^{(1)}(T)$ becomes

$$P_4^{(1)}(T) \sim P_2(T) + \frac{\pi^2}{16}\Delta_R^2 T - \frac{3}{8}\Delta_R^2 \omega_{eff}, \quad \text{as} \quad T \to \infty. \tag{8.6.20}$$

The integral $P_4^{(2)}(T)$ in the leading order in powers of $\beta \to 0$ becomes

$$P_4^{(2)}(T) = \int_0^{\omega_{eff}} d\omega\, \omega^2 \ln\left[1 - e^{-\beta\bar{\omega}}\right] \sim \int_0^{\omega_{eff}} d\omega\, \omega^2 \ln \beta\bar{\omega}, \quad \text{as} \quad \beta \to 0, \tag{8.6.21}$$

in accordance with the expansion (8.6.5), since the variable ω is restricted, and hence the variable $\bar{\omega} = \sqrt{\omega^2 + (a/2)^2}$ as well, where again $a = \sqrt{3}\Delta_R$. From the relations (7.5.3) and (7.5.8) it follows that $\omega' \to \bar{\omega}$ by $\Delta_R \to (1/2)\Delta_R$, so making this replacement in the expansion (8.6.10), one automatically obtains the high-temperature expansion for the integral $P_4^{(2)}(T)$ as follows:

$$P_4^{(2)} \sim \frac{1}{6}\omega_{eff}^3 \ln\left(\frac{\bar{\omega}_{eff}^2}{T^2}\right) - \frac{1}{9}\omega_{eff}^3 + \frac{1}{4}\Delta_R^2\omega_{eff} - \frac{\sqrt{3}}{8}\Delta_R^3 \arctan\left(\frac{2\omega_{eff}}{\sqrt{3}\Delta_R}\right), \tag{8.6.22}$$

where $\bar{\omega}_{eff}^2 = \omega_{eff}^2 + (3/4)\Delta_R^2$ as it was above.

The high-temperature expansion for the composition (8.5.4), on account of the relations (8.6.11) (8.6.12) and the previous expansions (8.6.10), (8.6.20), and (8.6.22) and after doing some algebra, becomes

$$\begin{aligned} \frac{16}{\pi^2}TM(T) \sim \frac{18}{\pi^2}\Delta_R^2\omega_{eff}T - \Delta_R^2T^2 + \frac{8}{3\pi^2}\omega_{eff}^3 T \ln\left(\frac{\omega_{eff}'^2}{\bar{\omega}_{eff}^2}\right) \\ + \frac{2\sqrt{3}}{\pi^2}\Delta_R^3 T \arctan\left(\frac{2\omega_{eff}}{\sqrt{3}\Delta_R}\right) - \frac{16\sqrt{3}}{\pi^2}\Delta_R^3 T \arctan\left(\frac{\omega_{eff}}{\sqrt{3}\Delta_R}\right), \end{aligned} \tag{8.6.23}$$

Substituting this expansion into equation (8.6.1), and on account of the expansion (8.6.6), we obtain in the $T \to \infty$ case

$$\begin{aligned} P_{NP}(T) \sim \frac{12}{\pi^2}\Delta_R^2\omega_{eff}T + \frac{8}{3\pi^2}\omega_{eff}^3 T \ln\left(\frac{\omega_{eff}'^2}{\bar{\omega}_{eff}^2}\right) \\ + \frac{2\sqrt{3}}{\pi^2}\Delta_R^3 T \arctan\left(\frac{2\omega_{eff}}{\sqrt{3}\Delta_R}\right) - \frac{16\sqrt{3}}{\pi^2}\Delta_R^3 T \arctan\left(\frac{\omega_{eff}}{\sqrt{3}\Delta_R}\right). \end{aligned} \tag{8.6.24}$$

One concludes that the exact cancelation of the $\Delta_R^2 T^2$ term occurs within the non-perturbative pressure itself. Thus $P_{NP}(T) \sim T$ to the leading order in the $T \to \infty$ limit. At the same time, at high temperatures the exact cancelation of the $P_{SB}(T)$ term occurs within the composition $(16/\pi^2)TM_1(T) = (16/\pi^2)T\,[P_2(T) - P_4(T)]$, which enters the composition (8.6.23). To show this explicitly, let us substitute into the former composition the relation (8.6.7), on account of the relations (8.6.8), and the relation (8.6.11), on account of the expansion (8.6.20), and doing some algebra, one obtains

$$\begin{aligned}
&\frac{16}{\pi^2}TM_1(T) \\
&\quad \sim -2P_{SB}(T) + 2P_{SB}(T) - \Delta_R^2 T^2 + \frac{6}{\pi^2}\Delta_R^2 \omega_{eff} T - \frac{16}{\pi^2}TP_4^{(2)}(T) \\
&\quad \sim -\Delta_R^2 T^2 + \frac{6}{\pi^2}\Delta_R^2 \omega_{eff} T - \frac{16}{\pi^2}TP_4^{(2)}(T), \quad \text{as} \quad T \to \infty,
\end{aligned} \tag{8.6.25}$$

from which the above-mentioned exact cancelation explicitly follows. The exact cancelation of the $P_2^{(2)}(T)$ terms and the expansion (8.6.22) for the $P_4^{(2)}(T)$ term are not shown, for simplicity.

Let us now consider equations (8.4.4), which is convenient to decompose as follows:

$$P_{PT}^s(T) = \frac{9}{2\pi^2}\alpha_s \Delta_R^2 \int_{\Lambda_{YM}}^{\infty} \mathrm{d}\omega\, \omega^2\, \frac{1}{\bar{\omega}} \frac{1}{e^{\beta\bar{\omega}} - 1} = P_1^s(T) - P_2^s(T), \tag{8.6.26}$$

where componants are

$$\begin{aligned}
P_1^s(T) &= \frac{9}{2\pi^2}\alpha_s \Delta_R^2 \int_0^{\infty} \mathrm{d}\omega\, \omega^2\, \frac{1}{\bar{\omega}} \frac{1}{e^{\beta\bar{\omega}} - 1}, \\
P_2^s(T) &= \frac{9}{2\pi^2}\alpha_s \Delta_R^2 \int_0^{\Lambda_{YM}} \mathrm{d}\omega\, \omega_R^2\, \frac{1}{\bar{\omega}} \frac{1}{e^{\beta\bar{\omega}} - 1},
\end{aligned} \tag{8.6.27}$$

and let us remind that $\bar{\omega} = \sqrt{\omega^2 + (3/4)\Delta_R^2} = \sqrt{\omega^2 + (a/2)^2}$.

In the integral $P_1^s(T)$ it is convenient to introduce a new dimensionless variable $x = \beta\bar{\omega} = \beta\sqrt{\omega^2 + (a/2)^2}$. After doing some algebra, it becomes

$$P_1^s(T) = \frac{9}{2\pi^2}\alpha_s\Delta_R^2 \int_0^\infty d\omega\, \omega^2\, \frac{1}{\bar{\omega}}\frac{1}{e^{\beta\bar{\omega}}-1}$$

$$= \frac{9}{2\pi^2}\alpha_s\Delta_R^2 T^2 \int_{\beta a/2}^\infty dx\, \frac{\sqrt{x^2-(\beta a/2)^2}}{e^x-1}. \tag{8.6.28}$$

The last integral at $\beta \to 0$ can be approximated to the leading order as,

$$\int_{(\beta a/2)}^\infty dx\, \frac{\sqrt{x^2-(\beta a/2)^2}}{e^x-1} \sim \int_0^\infty dx\, \frac{x}{e^x-1} = \frac{\pi^2}{6}, \tag{8.6.29}$$

then for the integral $P_1^s(T)$ to the leading order in powers of T we get

$$P_1^s(T) \sim \frac{3}{4}\alpha_s\Delta_R^2 T^2, \quad \text{as} \quad T \to \infty. \tag{8.6.30}$$

In the integral $P_2^s(T)$ the variable ω is restricted, and hence $\bar{\omega}$ as well. So to the leading order in the $T \to \infty$ limit this integral can be approximated,

$$P_2^s(T) = \frac{9}{2\pi^2}\alpha_s\Delta_R^2. \int_0^{\Lambda_{YM}} d\omega\, \omega^2\, \frac{1}{\bar{\omega}}\frac{1}{e^{\beta\bar{\omega}}-1}$$

$$\sim \frac{9}{2\pi^2}\alpha_s\Delta_R^2 T \int_0^{\Lambda_{YM}} d\omega\, \frac{\omega^2}{\bar{\omega}^2}, \tag{8.6.31}$$

in accordance with the expansion (8.6.5). The last formula can easily be integrated and thus the high-temperature expansion for the $P_2^s(T)$ term looks like

$$P_2^s(T) \sim \frac{9}{2\pi^2}\alpha_s\Delta_R^2 T \left[\Lambda_{YM} - \frac{\sqrt{3}}{2}\Delta_R \arctan\left(\frac{2\Lambda_{YM}}{\sqrt{3}\Delta_R}\right)\right]. \tag{8.6.32}$$

Summing it with the expansion (8.6.30), for the integral (8.6.26) one obtains in the $T \to \infty$ limit

$$P_{PT}^s(T) \sim \frac{9}{2\pi^2}\alpha_s\Delta_R^2 \left[\frac{\pi^2}{6}T^2 - T\left[\Lambda_{YM} - \frac{\sqrt{3}}{2}\Delta_R \arctan\left(\frac{2\Lambda_{YM}}{\sqrt{3}\Delta_R}\right)\right]\right], \tag{8.6.33}$$

which is nothing but the high-temperature expansion for the α_s-dependent perturbative part of the gluon pressure.

The high-temperature expansion of the gluon pressure is to be obtained by summing up the expansions (8.6.24) and (8.6.33), so it is

$$
\begin{aligned}
P_g(T) &= [P_{NP}(T) + P^s_{PT}(T)] \\
&\sim \frac{12}{\pi^2}\Delta_R^2 \omega_{eff} T + \frac{8}{3\pi^2}\omega_{eff}^3 T \ln\left(\frac{\omega'_{eff}}{\bar{\omega}_{eff}}\right)^2 \\
&+ \frac{2\sqrt{3}}{\pi^2}\Delta_R^3 T \arctan\left(\frac{2\omega_{eff}}{\sqrt{3}\Delta_R}\right) - \frac{16\sqrt{3}}{\pi^2}\Delta_R^3 T \arctan\left(\frac{\omega_{eff}}{\sqrt{3}\Delta_R}\right) \\
&+ \frac{9}{2\pi^2}\alpha_s \Delta_R^2 \left[\frac{\pi^2}{6}T^2 - T\left(\Lambda_{YM} - \frac{\sqrt{3}}{2}\Delta_R \arctan\left(\frac{2\Lambda_{YM}}{\sqrt{3}\Delta_R}\right)\right)\right].
\end{aligned}
\tag{8.6.34}
$$

Let us emphasize that the high-temperature expansions for all three pressures (8.6.24), (8.6.33), and (8.6.34) non-analytically depend on the mass gap in terms $\sim \Delta_R^3 T \sim (\Delta_R^2)^{(3/2)} T$, but it is not an expansion parameter like α_s. From asymptotics (8.6.33) and (8.6.34) it follows that $P^s_{PT}(T)$, and hence $P_g(T)$, behaves like T^2 to the leading order, while remembering $P_{NP}(T) \sim T$, see expansion (8.6.24). It is also interesting to note that the effective massive gluonic excitations $\omega'_{eff} = \sqrt{\omega_{eff}^2 + 3\Delta_R^2}$ and $\bar{\omega}_{eff} = \sqrt{\omega_{eff}^2 + (3/4)\Delta_R^2}$ are logarithmically suppressed at high temperatures, while there is no dependence on the effective massive gluonic excitation $\tilde{\omega}_{eff} = \sqrt{\Lambda_{YM}^2 + (3/4)\Delta_R^2}$ at all. In a more compact form the previous expansion looks like in the $T \to \infty$ limit,

$$
P_g(T) \sim B_2 \alpha_s \Delta_R^2 T^2 + \left[B_3 \Delta_R^3 + M^3\right] T, \tag{8.6.35}
$$

where M^3 denotes the terms of the dimensions of the GeV^3, which depend analytically on the mass gap Δ_R^2. The explicit expressions for it and for both constants B_2 and B_3 can be easily restored from the expansion (8.6.34), if necessary.

8.7 Discussion and conclusions

It is instructive to briefly discuss the asymptotic properties of all three pressures in more detail. Below T_c all pressures are exponentially suppressed in the low-temperature, $T \to 0$ limit, as plotted on Fig. 8.4.1. This is explicitly shown analytically in Section 8.5. The high-temperature, $T \to \infty$ expansion is explicitly evaluated in Section 8.6. At moderately high temperatures up to approximately $(3-4)T_c$ the exact functional dependence on

T remains rather complicated. It cannot be determined by the analytical evaluation of the integrals (7.5.12), (7.5.13), and (8.4.4) — only numerically as shown in Fig. 8.4.1. This non-trivial T-dependence can also be seen in Figs. 8.7.1, 8.7.2, and 8.7.3.

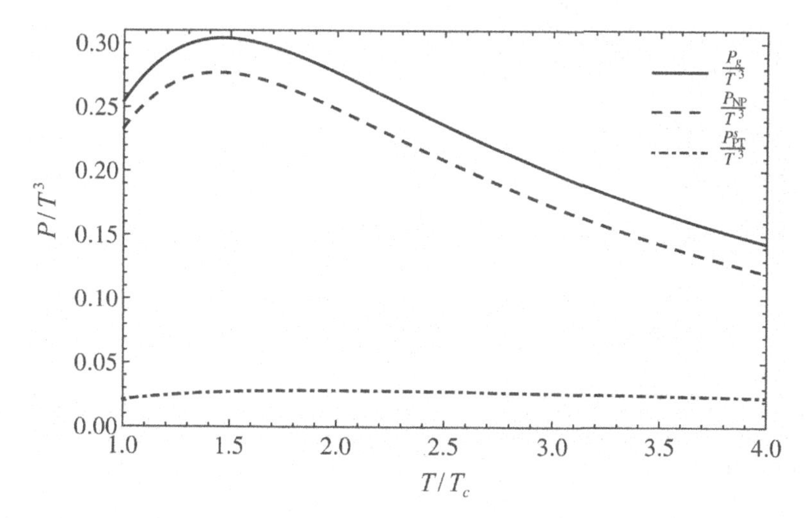

Fig. 8.7.1 The gluon pressure (8.4.3), the non-perturbative pressure (7.5.11) and the α_s-dependent perturbative pressure (8.4.4), all properly scaled in GeV units, are shown as functions of T/T_c.

In each of these figures all three pressures are scaled in the same way. Fig. 8.7.4 can be interpreted as clear diagrammatic evidence of the exact cancelation of the Stefan–Boltzmann terms rather close to T_c. It is analytically shown in equation (8.6.25) — it occurs within the composition $16TM_1(T)/\pi^2T^4$. In the non-perturbative pressure (7.5.11) the exact cancelation of the mass gap terms $\Delta_R^2T^2$ also occurs somewhere rather close to T_c, see equation (8.2.3). For the analytical evaluation of this phenomenon see Section 8.6, in general, and the high-temperature expansion (8.6.24), in particular. As a result, the non-perturbative pressure (7.5.11) will scale as T, while the perturbative pressure (8.4.4), and hence the gluon pressure (8.4.3), will continue to scale to the leading order as $\sim \alpha_s\Delta_R^2T^2$ in equations (8.6.33) and (8.6.34) respectively. Thus both pressures will approach the same constant in the limit of high temperature in Fig. 8.7.5, but very slowly since the term $\sim T^2$ is suppressed by the α_s order. At $T = 23T_c$ the non-perturbative pressure goes below the perturbative one, see Figs. 8.7.5 and 8.7.6. In general, all pressures are polynomials in integer

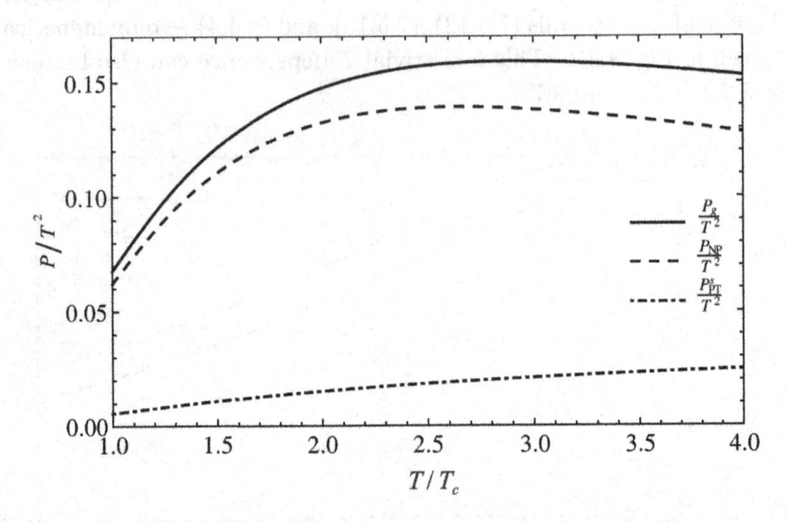

Fig. 8.7.2 The gluon pressure (8.4.3), the non-perturbative pressure (7.5.11) and the α_s-dependent perturbative pressure (8.4.4), all properly scaled in GeV^2 units, are shown as functions of T/T_c.

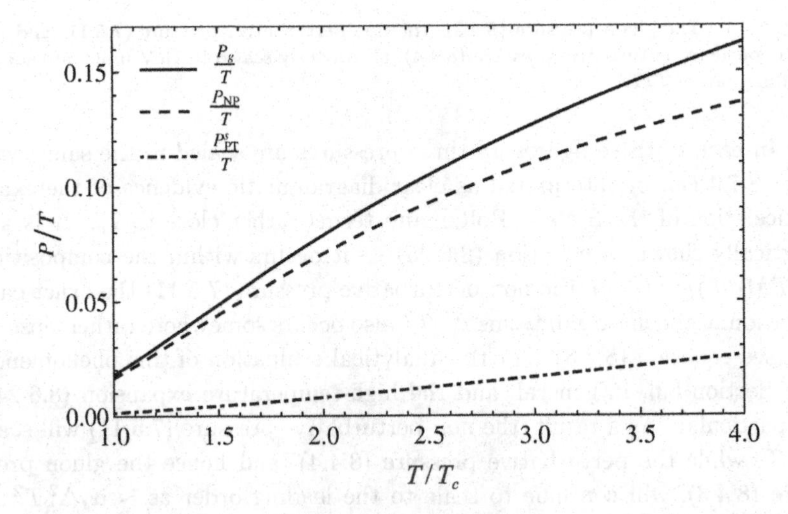

Fig. 8.7.3 The gluon pressure (8.4.3), the non-perturbative pressure (7.5.11) and the α_s-dependent perturbative pressure (8.4.4), all properly scaled in GeV^3 units, are shown as functions of T/T_c.

powers of T up to T^2 at very high temperatures. The term $\sim T^2$ has been first introduced in the phenomenological equation of state in Ref. [218] and in Refs. [219–223] as well. On the contrary, in our approach both terms $\sim T^2$ and $\sim T$ have not been introduced by hand. They naturally appear as a result of the explicit presence of the mass gap from the very beginning in the nonperturbative analytical equation of state [215].

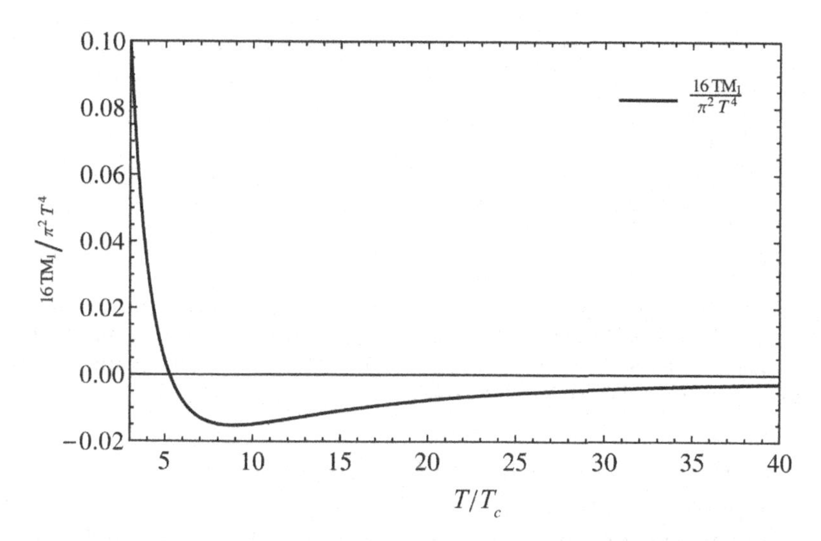

Fig. 8.7.4 The composition (8.6.25) scaled by T^4 is shown as a function of T/T_c. It approaches zero from below. This means that it does not contain the constant term — the Stefan – Boltzmann terms have already been exactly canceled far below $5T_c$.

Our final conclusions are collected as follows:

- We have developed the analytic thermal perturbation theory in the form of a convergent series, which made it possible to calculate the perturbative part of the gluon pressure termwise in integer powers of a small α_s.
- We have shown that the perturbative contribution of the α_s-order is numerically much smaller than the non-perturbative term in the range up to $23\,T_c$, see Figs. 8.4.1–8.7.3, 8.7.5, and 8.7.6.
- In the gluon pressure (8.4.1) the higher order terms in integer powers of a small α_s, which are determined by the convergent expansions (8.2.29) and (8.2.32), are numerically very small and therefore can be neglected.

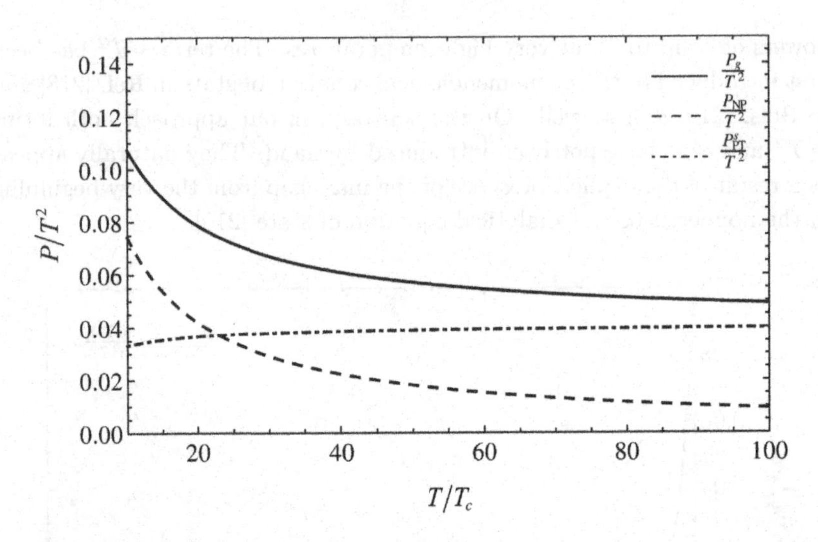

Fig. 8.7.5 The high temperature asymptotics of the gluon pressure (8.4.3), the non-perturbative pressure (7.5.11) and the α_s-dependent perturbative pressure (8.4.4) in GeV^2 units are shown as functions of T/T_c. At $T = 23T_c$ the non-perturbative pressure $P_{NP}(T)$ goes below the perturbative pressure $P^s_{PT}(T)$.

- All three pressures divided by $T^4/3$ have maxima at the temperature $T = T_c = 266.5$ MeV. Their low- (below T_c) and high-temperature (above T_c) expansions have been evaluated in Sections 8.5 and 8.6, respectively.
- In the low-temperature $T \to 0$ limit all three pressures are exponentially suppressed as in Fig. 8.4.1 due to the corresponding asymptotic of the gluon mean number. In the low-temperature expansion of the nonperturbative part a non-analytical dependence on the mass gap appears in terms $\sim (\Delta_R^2)^{1/2}T^3 \sim \Delta_R T^3$, see equation (8.5.28).
- The complicated mass gap- and T-dependence of all three pressures near T_c and up to approximately $(3-4)T_c$ is seen in Figs. 8.4.1, 8.7.1, 8.7.2, and 8.7.3. In particular, the fall off of each pressure is not exactly power-type. This clearly follows from Figs. 8.7.1, 8.7.2, and 8.7.3.
- The polynomial character of the high-temperature expansions for all three pressures is confirmed due to the corresponding asymptotic of the gluon mean number.

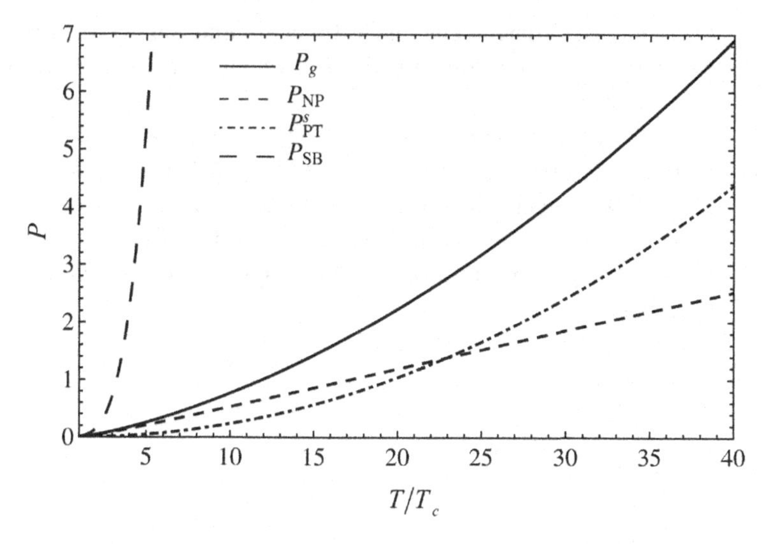

Fig. 8.7.6 The high temperature asymptotics of the gluon pressure (8.4.3), the nonperturbative pressure (7.5.11) and the α_s-dependent perturbative pressure (8.4.4) in GeV^4 units are shown as functions of T/T_c. At $T = 23T_c$ the pressure P_{NP} goes below $P^s_{PT}(T)$. The Stefan – Boltzmann pressure $P_{SB}(T) = (8/45)\pi^2 T^4$, formally extended up to zero temperature, is also shown.

- For the non-perturbative pressure (7.5.11) it contains only terms $\sim T$, and some of them depend non-analytically on the mass gap, namely $\sim (\Delta^2_R)^{3/2} T \sim \Delta^3_R T$.
- For the α_s-dependent perturbative contribution (8.4.4) it contains the terms $\sim T^2$ and $\sim T$ with a non-analytical dependence on the mass gap as above.
- For the gluon pressure (8.4.3) it contains both type of terms.

– In low- and high-temperature expansions, a non-analytical dependence on the mass gap occurs, as described above, but it is not an expansion parameter like α_s.
– In the gluon pressure (8.6.34) the mass gap term $\sim \Delta^2_R T^2$ is explicitly present in the whole temperature range, though it is suppressed by the α_s order.
– The perturbative part (8.6.26) dominates over its non-perturbative counterpart (8.6.1) in the limit of very high temperature starting from $T = 23T_c$, see Figs. 8.7.5 and 8.7.6. It is expected from the general point of view. This underlines once more the importance of the α_s-dependent perturbative pressure calculated here.

- Our analytical derivations in Sections 8.5 and 8.6 are in complete agreement with our numerical results shown in Figs. 8.4.1–8.7.6 and vice versa.
- In addition to the two massless ω_1, ω_2 and the two massive ω', $\bar{\omega}$ gluonic excitations described by $P_{NP}(T)$ (7.5.11), we have also one new massive excitation $\alpha_s \cdot \bar{\omega}$ described by $P^s_{PT}(T)$ (8.4.4). All these excitations are of the non-perturbative, dynamical origin, since in the perturbation theory $\Delta^2_R = 0$ limit they vanish from the spectrum.

Problems

Problem 8.1. *Derive the equation* (8.2.9).

Problem 8.2. *Perform the summation of the thermal logarithms in expression* (8.2.15) *in order to obtain equation* (8.2.16).

Problem 8.3. *Derive the equation* (8.2.23).

Problem 8.4. *Derive the expressions given by equations* (8.2.31) *and* (8.2.32).

Problem 8.5. *Repeat the derivations made in Section 8.5.*

Problem 8.6. *Repeat all the non-trivial derivations made in Section 8.6.*

Chapter 9

The Non-perturbative Analytical Equation of State for $SU(3)$ Gluon Plasma

9.1 Introduction

The main purpose of this Chapter is to complete the derivation of the non-perturbative analytical equation of state for the gluon matter. Its non-perturbative part which depends on the mass gap has been evaluated in Chapter 7. It determines the thermodynamic structure of the gluon matter at low temperatures. Its perturbative part together with the mass gap depends on the QCD fine-structure constant α_s, and it has been evaluated in Chapter 8. Here we are going to include the free massless gluons contribution or, equivalently, the Stefan – Boltzmann term to the non-perturbative analytical equation of state. This term determines the thermodynamic structure of the gluon matter at very high temperatures. Due to the initial normalization condition of the free perturbative vacuum to zero, this should be done in a rather sophisticated way, since it cannot be simply added to the equation of state at high temperatures. However, let us start by gathering all our previous analytical and numerical results. It is instructive to do this in order for the reader to have a general picture at hand.

9.2 The gluon pressure $P_g(T)$

In the gluon pressure (8.4.3), namely

$$P_g(T) = P_{NP}(T) + P_{PT}^s(T), \tag{9.2.1}$$

it is convenient to present the nonperturbative pressure $P_{NP}(T)$ in the equivalent form as follows:

$$P_{NP}(T) = \Delta_R^2 T^2 - \frac{6}{\pi^2}\Delta_R^2 P_1'(T) + \frac{16}{\pi^2} T M(T), \tag{9.2.2}$$

where two terms are

$$P_1'(T) = \int_0^{\omega_{eff}} d\omega \frac{\omega}{e^{\beta\omega} - 1} = - \int_{\omega_{eff}}^{\infty} d\omega \frac{\omega}{e^{\beta\omega} - 1} + \frac{\pi^2}{6} T^2, \tag{9.2.3}$$

and

$$M(T) = P_2(T) + P_3(T) - P_4(T), \tag{9.2.4}$$

while all the integrals $P_n(T)$, $n = 2, 3, 4$ are given as follows:

$$P_2(T) = \int_{\omega_{eff}}^{\infty} d\omega\, \omega^2 \ln\left[1 - e^{-\beta\omega}\right],$$

$$P_3(T) = \int_0^{\omega_{eff}} d\omega\, \omega^2 \ln\left[1 - e^{-\beta\omega'}\right],$$

$$P_4(T) = \int_0^{\infty} d\omega\, \omega^2 \ln\left[1 - e^{-\beta\bar{\omega}}\right], \tag{9.2.5}$$

where $\omega_{eff} = 1$ GeV and the mass gap $\Delta_R^2 = 0.4564$ GeV^2 for $SU(3)$ gauge theory have been fixed in the previous Chapter. ω' and $\bar{\omega}$ are then given by the relations

$$\omega' = \sqrt{\omega^2 + 3\Delta_R^2} = \sqrt{\omega^2 + m_{eff}'^2}, \quad m_{eff}' = \sqrt{3}\Delta_R = 1.17 \text{ GeV}, \tag{9.2.6}$$

and

$$\bar{\omega} = \sqrt{\omega^2 + \frac{3}{4}\Delta_R^2} = \sqrt{\omega^2 + \bar{m}_{eff}^2}, \quad \bar{m}_{eff} = \frac{\sqrt{3}}{2}\Delta_R = 0.585 \text{ GeV}, \tag{9.2.7}$$

respectively. It is worth reminding that in the non-perturbative pressure (9.2.2) $P_{NP}(T) = B_{YM}(T) + P_{YM}(T)$ the bag pressure $B_{YM}(T)$ (7.5.5) is responsible for the formation of the massive gluonic excitations ω' (9.2.6), while the Yang–Mills part $P_{YM}(T)$ (7.5.9) is responsible for the formation of the massive gluonic excitations $\bar{\omega}$ (9.2.7).

Let us also remind that the gluon mean number based on the definition given in Ref. [184]

$$N_g \equiv N_g(\beta, \omega) = \frac{1}{e^{\beta\omega} - 1}, \quad \beta = T^{-1}, \tag{9.2.8}$$

which appears in the integrals (9.2.4)–(9.2.5), describes the distribution and correlation of massless gluons in the gluon matter. Replacing ω by $\bar{\omega}$ and ω' we can consider the corresponding gluon mean numbers as describing the distribution and correlation of the corresponding massive gluonic excitations in the gluon matter, see integrals $P_3(T)$ and $P_4(T)$ in equation (9.2.5). They are of non-perturbative dynamical origin, since their masses are due to the mass gap Δ_R^2. All three different gluon mean numbers range continuously from zero to infinity as in Ref. [184]. We have the two different massless excitations, propagating in accordance with the integral (9.2.4) and the first of the integrals (9.2.5). However, they are not free, since in the $\Delta_R^2 = 0$ limit they vanish — the composition $P_2(T) + P_3(T) - P_4(T)$ becomes zero in this case. The non-perturbative pressure describes the two different massive and the two different massless gluonic excitations. The gluon mean numbers are closely related to the pressure. Its exponential suppression in the $T \to 0$ limit and the polynomial structure in the $T \to \infty$ limit are determined by the corresponding asymptotics of the gluon mean numbers. This has been explicitly shown in the previous Chapter.

The conventionally called perturbative term $P^s_{PT}(T)$ is

$$P^s_{PT}(T) \equiv P^s_{PT}\left(\Delta_R^2; T\right) = \alpha_s \cdot \frac{9}{2\pi^2}\Delta_R^2 \int_{\Lambda_{YM}}^{\infty} d\omega\, \omega^2\, \frac{1}{\bar{\omega}}\frac{1}{e^{\beta\bar{\omega}} - 1}. \qquad (9.2.9)$$

Here we note again that this term describes the propagation of the same massive gluonic excitations $\bar{\omega}$ (9.2.7) which are suppressed by the α_s-order. We can consider it as a new massive excitation in the gluon matter and denote it as $\alpha_s \cdot \bar{\omega}$. When the interaction is formally switched off by letting $\alpha_s = \Delta_R^2 = 0$, the composition $M(T)$ becomes zero as it follows from equations (9.2.5), and thus the gluon pressure (9.2.1) itself. This is due to the normalization condition of the free perturbative vacuum to zero, also valid at non-zero T, as emphasized above. The gluon pressure (9.2.1) has been calculated and discussed in the previous Chapter. It is shown in Fig. 9.2.1. Below T_c the gluon pressure is exponentially suppressed in the $T \to 0$ limit, which is related to the low-temperature asymptotics of the gluon mean number (9.2.8) — as mentioned above. It has a maximum at some characteristic temperature, $T_c = 266.5$ MeV. Close to T_c and at moderately high temperatures up to approximately $(3-4)T_c$ its exact functional dependence on the mass gap Δ_R^2 and the temperature T remains rather complicated. This means that the non-perturbative effects due to the mass gap are still important up to this temperature. The gluon pressure has a polynomial character in integer powers of T up to T^2 at high temperatures. Concluding, let us remind once more that the term $P^s_{PT}(\Delta_R^2; T)$ is, in fact, non-perturbative, depending on the mass gap Δ_R^2, propagation of which is only suppressed by the α_s-order. However, at very high temperatures it dominates over the non-perturbative term, as it has been explicitly shown in the previous Chapter.

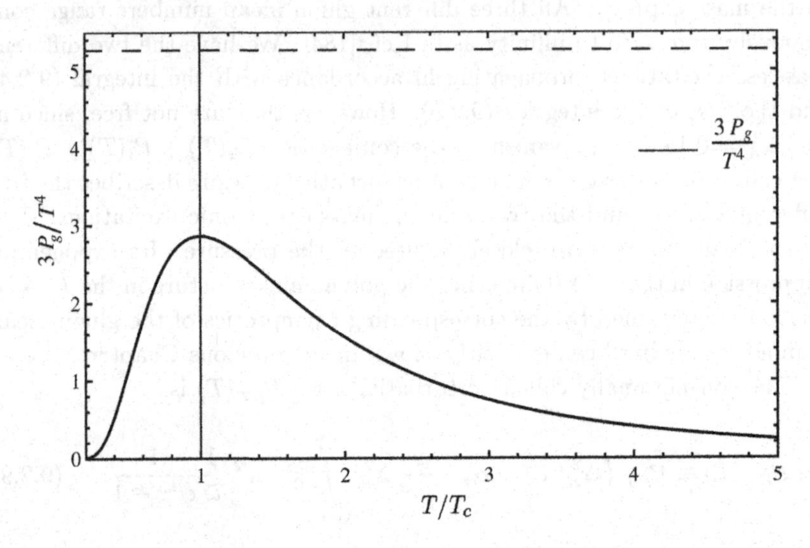

Fig. 9.2.1 The gluon pressure (9.2.1) divided by $T^4/3$ is shown as a function of T/T_c. It has a maximum at $T_c = 266.5$ MeV (*vertical solid line*). The *horizontal dashed line* is the general Stefan–Boltzmann constant $3P_{SB}(T)/T^4 = (24/45)\pi^2 \approx 5.2634$.

9.3 The full gluon plasma pressure

From Fig. 9.2.1 it clearly follows that the gluon pressure (9.2.1) will never reach the general Stefan–Boltzmann constant at high temperatures. That is not a surprise, since the Stefan–Boltzmann term has been canceled in the gluon pressure from the very beginning due to the normalization condition of the free perturbative vacuum to zero. There it has also been shown that the massless (but not free) gluons may be present at low temperatures, below T_c, in the gluon matter. However, their propagation in this region cannot be described by the Stefan–Boltzmann term itself. All this means that the Stefan–Boltzmann pressure has already been subtracted from the gluon pressure, but in a very specific way — the above-mentioned normalization condition is not simply its subtraction. The gluon pressure (9.2.1) may change its continuously falling off regime above T_c only in the near neighborhood of T_c in order for its full counterpart to reach the corresponding Stefan – Boltzmann limit at high temperatures. The Stefan – Boltzmann term is valid only at high temperatures; nevertheless, it cannot be added to equation (9.2.1) above T_c, even multiplied by the corresponding $\Theta((T/T_c) - 1)$-function. The problem is that in this case the pressure will get a jump at $T = T_c$, which is not acceptable. The full pressure is always a continuous growing function of temperature at any point of its domain. This means that we should add some other terms valid below T_c in order to restore a continuous character of the full pressure across T_c. This can be achieved by imposing a special continuity condition on these terms valid just at T_c. Moreover, the gluon pressure $P_g(T)$ itself should be multiplied by the function which is always negative below and above T_c. Otherwise, the full pressure will approach the Stefan – Boltzmann limit from above, which is not acceptable as well, since it should be approached in the asymptotic freedom way, slowly and from below. This term will also contribute to the condition of continuity for the full pressure. All these problems make the inclusion of the Stefan – Boltzmann term into the equation of state highly non-trivial. The most general form how this can be done is to add to equation (9.2.1) the term $\Theta((T/T_c) - 1)H(T) + \Theta((T_c/T - 1))L(T)$, valid in the whole temperature range, and the auxiliary functions $H(T)$ and $L(T)$ are to be expressed in terms of $P_{SB}(T)$ and $P_g(T)$.

The previous equations (9.2.1) then becomes

$$P_{GP}(T) = P_g(T) + \Theta\left(\frac{T_c}{T} - 1\right) L(T) + \Theta\left(\frac{T}{T_c} - 1\right) H(T), \qquad (9.3.1)$$

and its left-hand side here and below is denoted as $P_{GP}(T)$. This is the above-mentioned full counterpart, where subscript 'GP' means gluon plasma. In other words, after the inclusion of the Stefan – Boltzmann term into the equation of state the gluon matter will be called the gluon plasma. The gluon plasma pressure (9.3.1) is continuous at T_c if and only if

$$L(T_c) = H(T_c), \qquad (9.3.2)$$

which can be easily checked. Due to the continuity condition (9.3.2), the dependence on the corresponding Θ-functions disappears at T_c, and the gluon plasma pressure (9.3.1) remains continuous at any point of its domain. The role of the auxiliary function $L(T)$ is to change the behavior of $P_{GP}(T)$ from $P_g(T)$ at low, 'L' temperatures below T_c, especially in its near neighborhood, as well as to take into account the suppression of the Stefan – Boltzmann-type terms below T_c. The auxiliary function, $H(T)$ is aimed to change the behavior of $P_{GP}(T)$ from $P_g(T)$, as well as to introduce the Stefan–Boltzmann term itself and its modification due to asymptotic freedom at high, 'H' temperatures above and near T_c. These changes are necessary, since in the gluon pressure $P_g(T)$ the Stefan–Boltzmann term is missing, as described above, and it cannot be restored in a trivial way. So the appearance of the corresponding Θ-functions in the gluon plasma pressure (9.3.1) is inevitable together with the functions $H(T)$ and $L(T)$, playing only an auxiliary but still useful role from the technical point of view.

Analytical simulations

Actual analytical and numerical simulations are main subjects of this section. This need to be done in order to derive the non-perturbative analytical equation of state for the gluon plasma valid in the whole temperature range. Completing this program we will be able to compare our results and achieve the agreement with those of the thermal QCD lattice calculations in Refs. [221, 224]. The space of basic functions, in terms of which the auxiliary functions $L(T)$ and $H(T)$ should be found, has already been established. So on the general ground we can put

$$
\begin{aligned}
L(T) &= f_l(T)P_{SB}(T) - \phi_l(T)P_g(T), \\
H(T) &= f_h(T)P_{SB}(T) - \phi_h(T)P_g(T),
\end{aligned}
\tag{9.3.3}
$$

where all the dimensionless functions $f_l(T)$, $f_h(T)$ and $\phi_l(T)$, $\phi_h(T)$ will be called simulating functions. We call the functions $P_g(T)$ and $P_{SB}(T)$ as basic ones, since they are independent from each other and exactly known. They determine the structure of the gluon plasma pressure (9.3.1), while the simulating functions will mainly produce all the necessary corrections to their corresponding asymptotics at low and high temperatures as well as at T_c. This also makes it possible to use in what follows the exact relations, which result from our calculations shown in Fig. 9.2.1, namely

$$
\begin{aligned}
&[P_{SB}(T) - 1.84P_g(T)]_{T=T_c} = 0, \\
&\frac{\partial}{\partial T}[P_{SB}(T) - 1.84P_g(T)]_{T=T_c} = 0.
\end{aligned}
\tag{9.3.4}
$$

The Stefan–Boltzmann term $P_{SB}(T)$ may only appear below T_c if it is exponentially suppressed. This has to be also true for $P_g(T)$, since we need no additional gluon pressure in the $T \to 0$ limit. As we already know from previous Chapters, we will achieve this goal by choosing the simulating functions $f_l(T)$ and $\phi_l(T)$ due to the asymptotic behavior of the corresponding gluon mean number (9.2.8) in the $T \to 0$ limit. So putting them as functions of (T_c/T) to the leading order (i.e., neglecting the functional dependence on the terms $\sim (T/T_c)^{\lambda_i}$, where $\lambda_i > 0$), one obtains

$$f_l(T) = \textstyle\sum_{i=1}^{n} A_i e^{-\alpha_i (T_c/T)}, \qquad \text{where} \quad \alpha_i > 0, \quad \text{and}$$
$$\phi_l(T) = e^{-\alpha (T_c/T)}, \qquad \text{where} \quad \alpha > 0. \tag{9.3.5}$$

Here α and α_i are arbitrary positive numbers — in other words, we measure ω in equation (9.3.5) in terms of T_c with the help of these numbers. The constants A_i are arbitrary ones at this stage. The similar constant in the relation (9.3.5) for $\phi_l(T)$ is put to one without losing generality, since we need only suppression of the additional $P_g(T)$ in the $T \to 0$ limit, as underlined above, while at $T = T_c$ its additional contribution can be controlled by the parameter α only. From the gluon plasma pressure (9.3.1) below T_c and relations (9.3.3) and (9.3.5) it follows

$$\begin{aligned} P_{GP}(T) &= P_g(T) + f_l(T) P_{SB}(T) - \phi_l(T) P_g(T) \\ &\sim \left[1 - e^{-\alpha (T_c/T)}\right] P_g(T) + \sum_{i=1}^{n} A_i e^{-\alpha_i (T_c/T)} P_{SB}(T) \end{aligned} \tag{9.3.6}$$

in the $T \to 0$ limit. The additional contributions are indeed exponentially suppressed, and the asymptotic of the gluon plasma pressure $P_{GP}(T)$ is determined by the gluon pressure $P_g(T)$, as it should be. In this connection, let us note that $P_g(T) \sim T$ to the leading order in the $T \to 0$ limit, while $P_{SB}(T) \sim T^4$ (see the previous Chapter). At the same time, the corresponding gluon mean numbers (9.3.5) allow us to change the value of the gluon plasma pressure from the gluon pressure near T_c, as it is expected.

For the simulating function $f_h(T)$ our general choice is

$$f_h(T) = 1 - \alpha_s(T) = 1 - \frac{\alpha_s \nu_1}{1 + \alpha_s \nu_2 \ln(T/T_c)}, \qquad \text{while} \quad T \geq T_c, \tag{9.3.7}$$

where $\nu_1 > 0$ and $\nu_2 > 0$ are arbitrary real numbers at this stage.

Evidently, the first term leads to the correct Stefan–Boltzmann limit for the gluon plasma pressure (9.3.1). The second term in this expression mimics the perturbative formula for the effective charge (7.2.8) as a function of temperature. In applications at finite temperature, the ratio q^2/Λ_{YM}^2 in equation (7.2.8) is effectively replaced by the ratio T/T_c. That is the reason for the appearance of the numbers ν_1 and ν_2 in equation (9.3.7). In accordance with Ref. [193] we replaced the superscript 'PT' used in equation (7.2.8) by the subscript 's' in equation (9.3.7) for $\alpha_s(T)$. In other words, the Stefan – Boltzmann limit at high temperatures should be reached in the asymptotic freedom way — due to the relation (9.3.7). This should be true for any other independent thermodynamic quantity, such as the energy and entropy densities, etc. There are different empirical expressions for $\alpha_s(T)$ in Refs. [184, 191, 193, 199, 213, 216, 225]. However, any such expression can be re-calculated at any given value of T_c, and thus relate the different approaches for $\alpha_s(T)$ to each other. Here it is convenient to choose the expression for $\alpha_s(T)$ as it is given in Ref. [193] at $T_c = 172$ MeV. Re-calculated at $T_c = 266.5$ MeV, it becomes

$$f_h(T) = 1 - \alpha_s(T) = 1 - \frac{0.36}{1 + 0.55 \ln(T/T_c)}, \quad \text{while} \quad T \geq T_c, \qquad (9.3.8)$$

so it is uniquely fixed and in fact, this function is not the simulating one. Let us only note that the value of $\alpha_s = 0.1184$ is hidden in the numbers 0.36 and 0.55, thus, it can be restored as follows: $0.36 = 3.04\alpha_s$ and $0.55 = 4.64\alpha_s$, as it should be due to equation (9.3.7).

The simulating function $\phi_h(T)$ has again to be chosen in the form of the corresponding gluon mean number (9.2.8), but its asymptotic has to be taken in the $T \to \infty$ limit. It should be a regular function of T as it goes to infinity in order not to contradict the asymptotic of $P_g(T)$ in this limit. The asymptotic of the gluon plasma pressure (9.3.1) at high temperature $T \to \infty$ thus becomes

$$\begin{aligned} P_{GP}(T) &= P_g(T) + f_h(T)P_{SB}(T) - \phi_h(T)P_g(T) \\ &\sim [1 - \alpha_s(T)]\, P_{SB}(T) + [1 - \phi_h(T)]\, P_g(T), \end{aligned} \qquad (9.3.9)$$

where $[1 - \phi_h(T)]$ has to be negative above T_c, so the gluon plasma pressure will approach the term $[1 - \alpha_s(T)]\, P_{SB}(T)$ from below at high temperatures, as it is required. At the same time, the function $\phi_h(T)$ allows us to change the value of the gluon plasma pressure (9.3.1) from the gluon pressure $P_g(T)$ near T_c, as it is expected. Thus we have a general restriction, namely

$$1 - \phi_h(T) < 0, \quad \text{while} \quad T \geq T_c. \qquad (9.3.10)$$

The explicit expressions for the auxiliary functions $L(T)$ and $H(T)$ (9.3.3), via the chosen simulating functions (9.3.5), are

$$L(T) = \sum_{i=1}^{n} A_i e^{-\alpha_i (T_c/T)} P_{SB}(T) - e^{-\alpha(T_c/T)} P_g(T), \qquad (9.3.11)$$

and

$$H(T) = (1 - \alpha_s(T))\, P_{SB}(T) - \phi_h(T)P_g(T), \tag{9.3.12}$$

where $1 - \alpha_s(T)$ is determined by equation (9.3.8).

The gluon plasma pressure (9.3.1), on account of the relations (9.3.11) and (9.3.12), then looks like

$$\begin{aligned} P_{GP}(T) &= P_g(T) \\ &\quad + \Theta\left(\frac{T_c}{T} - 1\right)\left[\sum_{i=1}^{n} A_i e^{-\alpha_i (T_c/T)} P_{SB}(T) - e^{-\alpha(T_c/T)} P_g(T)\right] \\ &\quad + \Theta\left(\frac{T}{T_c} - 1\right)\left[(1 - \alpha_s(T))\, P_{SB}(T) - \phi_h(T)P_g(T)\right], \end{aligned} \tag{9.3.13}$$

so that from now on we can forget about the auxiliary functions $L(T)$ and $H(T)$, though they are still useful from the technical point of view in the analytical evaluation of various thermodynamic quantities.

From the relations (9.3.11), (9.3.12), and using the relations (9.3.4), it follows that at $T = T_c$ the relation (9.3.2) becomes

$$1.84 \cdot \sum_{i=1}^{n} A_i e^{-\alpha_i} - e^{-\alpha} = 1.84 \cdot (1 - 0.36) - \phi_h(T_c) = 1.1776 - \phi_h(T_c). \tag{9.3.14}$$

Due to this relation, from the previous expression (9.3.13) at $T = T_c$, one obtains

$$\begin{aligned} P_{GP}(T_c) &= P_g(T_c) + [1.1776 - \phi_h(T_c)]P_g(T_c) \\ &= P_g(T_c) + \left[1.84 \sum_{i=1}^{n} A_i e^{-\alpha_i} - e^{-\alpha}\right] P_g(T_c), \end{aligned} \tag{9.3.15}$$

which shows that it depends on the number $\phi_h(T_c)$ only, since the value $P_g(T_c)$ is known from our calculations (see, for example Fig. 9.2.1). From the expression (9.3.15) it also follows that

$$2.1776 - \phi_h(T_c) \geq 0, \tag{9.3.16}$$

since the full pressure (9.3.15) should be always positive, in particular at the critical temperature, T_c.

Analytical and numerical simulation of the gluon plasma pressure above T_c

Our aim here is to find the simulating function $\phi_h(T)$ by fitting lattice data at high temperatures above T_c. For this, let us derive from the gluon plasma equation of state (9.3.13) its values at $a = T/T_c$, $a \geq 1$ as follows:

$$P_{GP}(a) = f_h(a)P_{SB}(a) + [1 - \phi_h(a)]\, P_g(a), \qquad (9.3.17)$$

and $f_h(a)$ is given in equation (9.3.8), namely

$$f_h(a) = 1 - \alpha_s(a) = \left(1 - \frac{0.36}{1 + 0.55 \ln a}\right). \qquad (9.3.18)$$

Adjusting our parametrization of the gluon plasma pressure (9.3.17) to that used in recent lattice simulations for the Yang–Mills $SU(3)$ case at $T = aT_c$ in Ref. [221], one obtains

$$\frac{3P_{GP}(T)}{T^4} = \frac{P_l(T)}{T^4} \cdot 5.2634, \qquad (9.3.19)$$

where $5.2634 = 3P_{SB}(T)/T^4$ is the general Stefan–Boltzmann constant shown up to four digits only after the point. For simplicity, we will keep this for all our numbers for the following. At the same time, let us stress that we have carried out all our calculations keeping six digits after the point — thus we can calculate all numbers to any requested accuracy. However, where it will not cause a big uncertainty we will show one or two digits after the point. The subscript 'l' in $P_l(T)/T^4$ is due to the above-mentioned lattice data, which, for example, should read $P_l(T_c)/T_c^4 = 0.019676$, $P_l(2T_c)/(2T_c)^4 = 0.613278$, $P_l(3T_c)/(3T_c)^4 = 0.731751$ etc.

As we already know, the best way to choose the appropriate expression for the simulating function $\phi_h(a)$ is to mimic again the asymptotic of the gluon mean number (9.2.8) but in the $T \to \infty$ limit, which is equivalent to the $a \to \infty$ limit. As a function of a, one can write

$$\phi_h(a) = \frac{a^{-1}}{e^{(\mu/a)} - 1} = \frac{a^{-1}}{\sum_{k=1}^{\infty} \frac{1}{k!}\left(\frac{\mu}{a}\right)^k} = \frac{1}{\sum_{k=0}^{\infty} c'_{k+1} a^{-k}} = \sum_{k=0}^{\infty} c_k a^{-k}, \qquad (9.3.20)$$

where $c'_k = \frac{1}{k!}\mu^k$ and $c_0 c'_1 = 1$, $c_n + (1/c'_1) \sum_{k=1}^{n} c_{n-k} c'_{k+1} = 0$, $n = 1, 2, 3....$ So this simulating function at high temperatures becomes a series in inverse powers of $a = T/T_c$, starting from non-zero $c_0 = \mu^{-1} > 1$ in agreement with the general restriction (9.3.10). This was the reason for the multiplication of the corresponding gluon mean number in equation (9.3.20) by a^{-1}. A possible arbitrary constant to which the initial a^{-1} has to be multiplied is set to one without loosing generality due to the arbitrariness of the constants c_k in the initial expansion (9.3.20) at this stage.

First of all, we are interested in

$$\phi_h(1) = \sum_{k=0}^{\infty} c_k, \tag{9.3.21}$$

as it follows from the previous equation (9.3.20). This important quantity appears in equations (9.3.14) and (9.3.15), since $\phi_h(1) \equiv \phi_h(T_c)$. On the other hand, the series (9.3.20) can be re-written as follows:

$$\begin{aligned} \phi_h(a) &= \sum_{k=0}^{\infty} c_k a^{-k} = \sum_{k=0}^{m} c_k a^{-k} + \sum_{k=m+1}^{\infty} c_k a^{-k} \\ &= \nu_0(a;m) + c_{m+1} a^{-m-1} + \sum_{k=m+2}^{\infty} c_k a^{-k}, \end{aligned} \tag{9.3.22}$$

where

$$\nu_0(a;m) = \sum_{k=0}^{m} c_k a^{-k} = c_0 + \mathcal{O}(a^{-1}), \quad \text{while} \quad a \to \infty. \tag{9.3.23}$$

At $a = 1$ the series at equation (9.3.22) yields

$$c_{m+1} = [\phi_h(1) - \nu_0(1;m)] - \sum_{k=m+2}^{\infty} c_k. \tag{9.3.24}$$

Substituting it back to the previous equation (9.3.22), one obtains

$$\begin{aligned} \phi_h(a) = &\ \nu_0(a;m) + [\phi_h(1) - \nu_0(1;m)]\, a^{-m-1} \\ &-a^{-m-1} \sum_{k=m+2}^{\infty} c_k + \sum_{k=m+2}^{\infty} c_k a^{-k}. \end{aligned} \tag{9.3.25}$$

This means that the whole expansion (9.3.20) or, equivalently the above (9.3.25) can be effectively approximated by a two-term polynomial,

$$\phi_h(a) = c_0 + [\phi_h(1) - c_0]\, a^{-m-1}, \tag{9.3.26}$$

which has the correct limit when a goes to infinity, equation (9.3.23), and it is self-consistent at $a = 1$. In other words, in both limits $a = [1, +\infty)$ it behaves like the initial infinite series (9.3.25). For future purpose it is convenient to present this equation in the equivalent form, namely

$$\phi_h(a) = c_0 \left[1 - a^{-n}\right] + c a^{-n}, \tag{9.3.27}$$

where we have put $n = m+1$ and $c = \phi_h(1)$ in comparison with the previous equation. Let us underline that all the three parameters c_0, $c = \phi_h(1)$, and $n > 0$ are independent from each other, remain arbitrary at this stage.

From the relation given by equation (9.3.16) follows the restriction, namely $\phi_h(1) = \phi_h(T_c) \leq 2.1776$. From the relation (9.3.10) we also know that $1 - c_0 < [\phi_h(a) - c_0]a^{-n}$ at any a. So at $a = 1$ then it follows that $1 < \phi_h(1)$, while when a goes to infinity this will be guaranteed if $1 - c_0 < 0$ or, equivalently, $c_0 > 1$ itself. All the numbers in the relation (9.3.26) or, equivalently, (9.3.27) are positive. In all these restrictions the equality is only a convention, since the pressure (9.3.15) cannot be exactly zero at T_c, which implies that $\phi_h(T_c) = 2.1776$. It is convenient to present both restrictions together as follows:

$$\begin{aligned} &1 < c_0, \\ &1 < c = \phi_h(1) \leq 2.1776. \end{aligned} \tag{9.3.28}$$

The fit to lattice data available from the high temperature interval in Ref. [221], namely $a = 2.800744 - 3.436657$ (which includes 33 data points), is to be performed with the help of the following equation

$$\frac{P_l(T)}{T^4} \cdot 5.2634 = 5.2634 f_h(a) + [1 - \phi_h(a)] \frac{3P_g(T)}{T^4}, \tag{9.3.29}$$

where $T = aT_c$ and the values of $P_l(T)/T^4$ have been taken from lattice data [221] as described above, while the values of $3P_g(T)/T^4$ have been taken from our data, Fig. 9.2.1. $f_h(a)$ and $\phi_h(a)$ are given in equations (9.3.18) and (9.3.27), respectively. Best fit has been achieved at

$$c_0 = 1.2815, \quad c = 2.1164, \quad n = 3 \tag{9.3.30}$$

by using the Least Mean Squares (LMS) method in Ref. [226], see Fig. 9.3.1. Let us underline that according to this method the solution for these parameters c_0, c and n is a unique one, satisfying the general restrictions (9.3.18). It is worth emphasizing that the average deviation is minimal at the values (9.3.30). Details of our calculations are briefly described in Appendix 9.C.

Hence the relation (9.3.27) becomes

$$\phi_h(a) = 1.2815 + 0.8349a^{-3}, \quad \phi_h(1) = 2.1164, \tag{9.3.31}$$

and as a function of T it finally becomes

$$\phi_h(T) = 1.2815 + 0.8349 \left(\frac{T_c}{T}\right)^3, \quad \phi_h(T_c) = 2.1164, \tag{9.3.32}$$

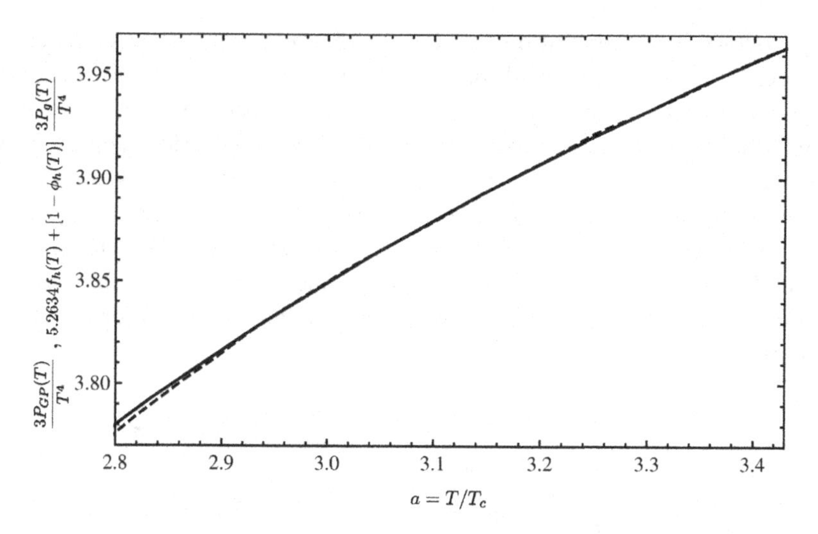

Fig. 9.3.1 The gluon plasma pressure as the right-hand-side of equation (9.3.33), as a function of a in the interval $a = 2.800744 - 3.436657$ is shown for $\phi_h(a)$ determined by the relation (9.3.32) (*solid line*). The lattice curve is taken from Ref. [221] for $SU(3)$ pressure. The left-hand side of the equation (9.3.33) is also shown (*dashed line*). Both lines are calculated in the same units (9.3.19).

determining this function up to the leading and next-to-leading orders in the $T \to \infty$ limit. Thus our method makes it possible to establish the behavior of the gluon plasma pressure in the temperature range $a \geq 1$ as well. This is to be achieved by the help of the following equation

$$\frac{3P_{GP}(T)}{T^4} = 5.2634 f_h(T) + [1 - \phi_h(T)]\ \frac{3P_g(T)}{T^4}, \tag{9.3.33}$$

where $\phi_h(T)$ is already known and given in equation (9.3.32), while $f_h(T)$ is explicitly given in equation (9.3.8) or, equivalently, in (9.3.18), see Fig. 9.3.2. It is instructive to explicitly show the value of the gluon pressure at T_c, which is

$$\frac{3P_{GP}(T_c)}{T_c^4} = 5.2634 \cdot 0.64 - 1.1164 \cdot \frac{3P_g(T_c)}{T_c^4} = 0.1738, \tag{9.3.34}$$

since $3P_g(T_c)/T_c^4 = 2.8616$, as it follows from our calculations. Let us remind that the lattice value at this point is

$$\frac{P_l(T_c)}{T_c^4} \cdot 5.2634 = 0.0196 \cdot 5.2634 = 0.1036, \tag{9.3.35}$$

as it follows from the relation (9.3.19). Taking into account the completely different nature of analytical and lattice approaches to QCD at non-zero temperature these numbers are surprisingly close to each other — it is worth reminding that we have adjusted our values to those of lattice ones far away from T_c. It is quite possible that a small deviation (up to the second digit after the point) of analytical curve from the lattice one in Fig. 9.3.2 has to be traced back to the details of the approximation scheme for the infinite series (9.3.20). We consider such a coincidence as the confirmation of our analytical calculations by lattice results and vice versa for $a \geq 1$ or, equivalently, $T \geq T_c$.

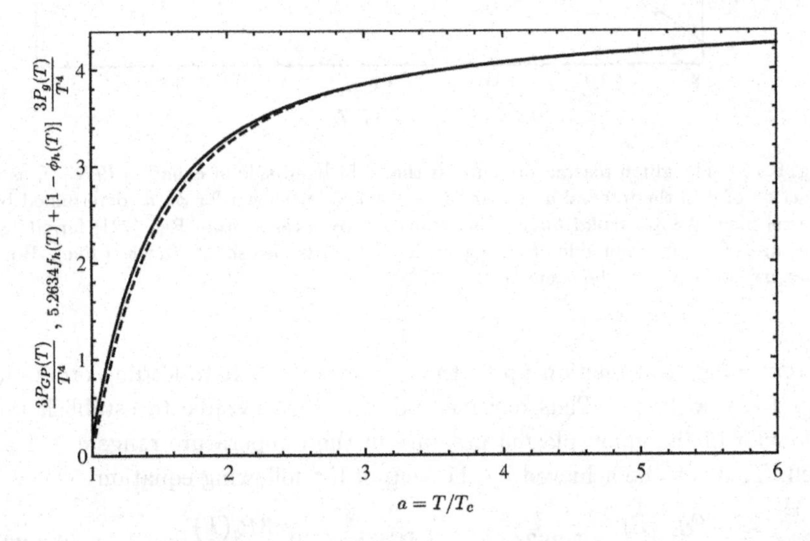

Fig. 9.3.2 The gluon plasma pressure (9.3.33) as a function of $a \geq 1$ is shown for $\phi_h(a)$ determined by the relation (9.3.32) (*solid curve*). The lattice curve based on Ref. [221] for $SU(3)$ pressure is also shown (*dashed curve*). Both curves are calculated in the same units (9.3.19).

Analytical and numerical simulation of the gluon plasma pressure below T_c

For future purpose it is convenient to begin this subsection with showing explicitly that the relation (9.3.14), on account of the value (9.3.32) for $\phi_h(T_c)$, becomes equivalent to

$$\sum_{i=1}^{n} A_i e^{-\alpha_i} = \left[0.5435e^{-\alpha} - 0.5102\right]. \qquad (9.3.36)$$

It makes it possible to reduce the number of independent parameters from $2n+1$ to $2n$, and obviously it has been derived at $a = 1$.

The fit to lattice data available from the low temperature interval in Ref. [221], namely $a = 0.907850 - 0.989528$ has been performed in Appendix 9.D. This made it possible to restore the true lattice pressure below T_c up to zero. However, here we cannot use lattice data just below T_c, since our value and the lattice one at T_c are slightly different, see equations (9.3.34) and (9.3.35). This means that we cannot reproduce almost identically the lattice pressure just below T_c. At the same time, this can be done at sufficiently low temperatures. The above-described fit has to be achieved with the help of the following equation

$$\frac{3P_{GP}(T)}{T^4} = 5.2634 \cdot \sum_{i=1}^{n} A_i e^{-\alpha_i (T_c/T)} + \left[1 - e^{-\alpha(T_c/T)}\right] \frac{3P_g(T)}{T^4}. \qquad (9.3.37)$$

Our best values for the parameters involved are:

$$\alpha = 0.0001, \quad \alpha_1 = 19.1,\ \alpha_2 = 3.4,\ A_1 e^{-\alpha_1} = 0.0282,\ A_2 e^{-\alpha_2} = 0.005. \qquad (9.3.38)$$

For convenience, this fit is shown in Fig. 9.3.3 up to $0.2T_c$ only. On the one hand, these values show that the deep penetration of the Stefan–Boltzmann-type term below T_c is indeed strongly suppressed, as expected. On the other hand, they satisfy the relation (9.3.36) as it should be. Perhaps the fit can be slightly improved by taking into account more terms in the left-hand-side of the relation (9.3.36). We restrict ourselves to the two terms only, since the values of the lattice pressure below T_c are very small and there are no convincing lattice data points below $0.9T_c$. Let us also note that we do not use the Least Mean Squares method here, since we have encountered some numerical problems with its non-linear realization. However, our fit made by 'hand' is very accurate for the interval starting below $0.8T_c$, see Fig. 9.3.3. Even above this value the difference between these two curves at the same temperature is very small — it appears in the second digit after the point, as expected.

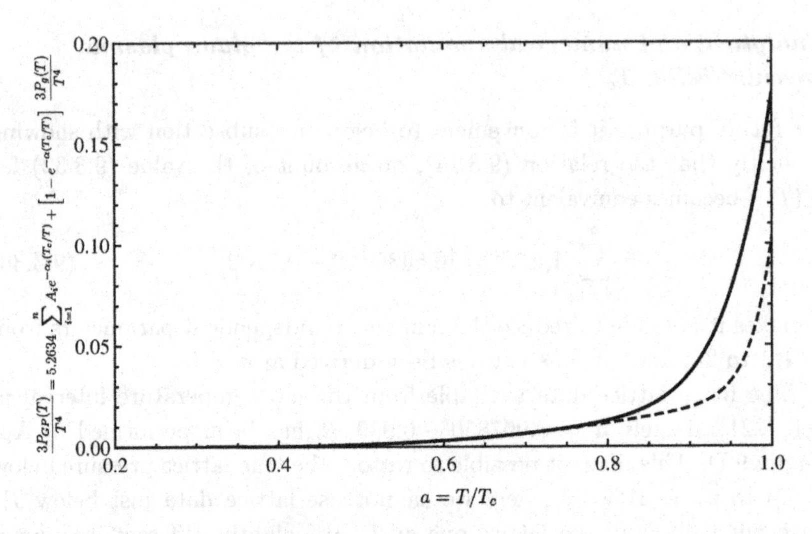

Fig. 9.3.3 The gluon plasma pressure (9.3.37) as a function of $a \leq 1$ is shown as a solid curve. The restored lattice curve for $SU(3)$ pressure is also shown (*dashed curve*). Starting from $0.8T_c$ they become almost identical. Both curves are calculated in the same units (9.3.19).

The gluon plasma pressure in the whole temperature range

The gluon plasma pressure is given by (9.3.13) in the whole temperature range $T = [0, +\infty)$, this finally becomes

$$\begin{aligned} P_{GP}(T) &= P_g(T) \\ &+ \Theta\left(\frac{T_c}{T} - 1\right)\left[0.0282 e^{-\alpha_1(\frac{T_c}{T}-1)} + 0.005 e^{-\alpha_2(\frac{T_c}{T}-1)}\right] P_{SB}(T) \\ &- \Theta\left(\frac{T_c}{T} - 1\right) e^{-\alpha(T_c/T)} P_g(T) \\ &+ \Theta\left(\frac{T}{T_c} - 1\right)\left[(1 - \alpha_s(T)) P_{SB}(T) - \phi_h(T) P_g(T)\right], \end{aligned} \tag{9.3.39}$$

where

$$\alpha_s(T) = \frac{0.36}{1 + 0.55\ln(T/T_c)}, \quad \text{while} \quad T \geq T_c, \tag{9.3.40}$$

and

$$\phi_h(T) = 1.2815 + 0.8349\left(\frac{T_c}{T}\right)^3, \quad \phi_h(T_c) = 2.1164. \tag{9.3.41}$$

So the gluon plasma pressure (9.3.39) is completely known now, since $P_g(T)$ is also exactly known, equation (9.2.1), while the numbers α_1, α_2 and α are present in the relation (9.3.38). It is shown in Figs. 9.3.3 and 9.3.4. This means that the auxiliary functions $L(T)$ and $H(T)$ in equation (9.3.1) have finally been fixed in terms of the basic functions $P_{SB}(T)$ and $P_g(T)$ within our approach. For simplicity, in what follows we will omit the subscript 'GP' in the gluon plasma pressure (9.3.39), as we will put $P_{GP}(T) \equiv P(T)$. The same will be done in the notations of all other thermodynamic quantities as well.

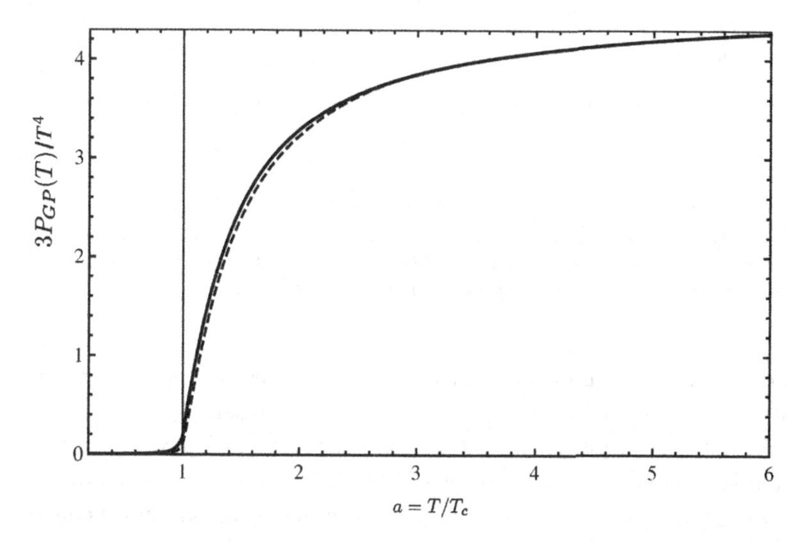

Fig. 9.3.4 The properly scaled gluon plasma pressure (9.3.39) is shown as a function of $a = T/T_c$ (*solid curve*). The lattice curve based on Ref. [221] for $SU(3)$ pressure is also shown (*dashed curve*). Both pressures are scaled in the same way in accordance with equation (9.3.19).

A few general remarks are in order. The requirement that the pressure is growing continuously (class C^0) as a function of temperature in the whole range is a rather strong restriction. This means that it is differentiable at any point of its domain, while at T_c the derivative itself may not be continuous — it may have a discontinuity at this point. At the same time, the adjustment of both terms at T_c, associated with the corresponding Θ-functions, has to be done in the above-mentioned requested way, thus the pressure should be a continuous function of the temperature across a possible phase transition at T_c as plotted on Figs. 9.3.4 and 9.3.5. In

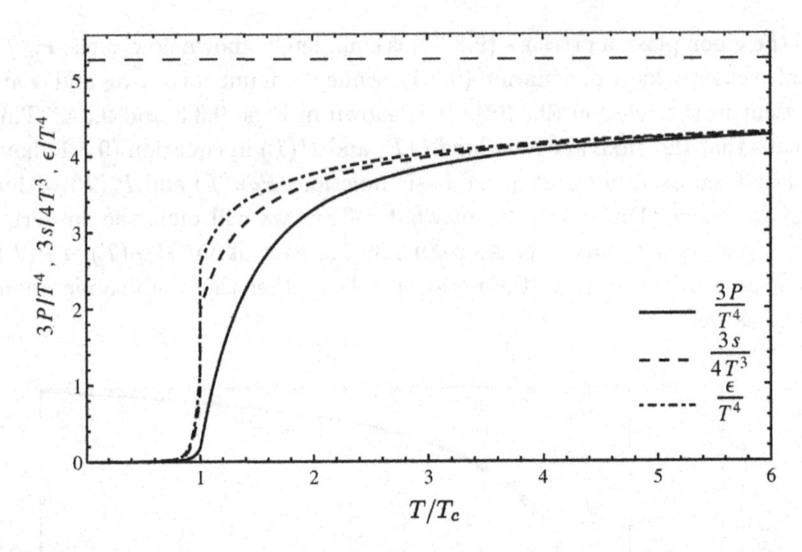

Fig. 9.3.5 The gluon plasma pressure (9.3.39), the entropy, and energy densities (9.4.1) all properly scaled are shown as functions of T/T_c. The finite jumps in densities are clearly seen, and the LH is $\epsilon_{LH} = 1.414$. Their common Stefan – Boltzmann limit (9.5.1) at high temperatures (*straight dashed line*) is rather slowly approaching.

principle, there are no other constraints on the parameters α_i, A_i and α apart from that all the derivatives of the gluon plasma pressure (9.3.39) should not gain negative values at low temperatures below T_c, but, they should exponentially approach zero from above.

Our only problem was how to restore the free massless gluons contribution to the full pressure $P_{GP}(T)$ (9.3.1), maintaining its continuous character across T_c. That is why we use only one type of the gluon mean numbers for each simulating function, namely the low-temperature asymptotic of their free massless type (9.2.8) in the form of the sum with different α_i and A_i parameters, equation (9.3.5). It is the general one for the simulating function $f_l(T)$ to the leading order, while for the simulating function $\phi_l(T)$ the chosen expression in equation (9.3.5) is fully sufficient, as explained above. It is worth emphasizing that the values of $P_{GP}(T)$ below T_c are rather small, so the approximation of the simulating functions $f_l(T)$ and $\phi_l(T)$ by their respective expressions to the leading order is justified. For the high-temperature asymptotic the resulting sum (9.3.20) for the simulating function $\phi_h(T)$ is the general one, of course, even using the sum with different μ parameters. On the one hand, this makes it possible to achieve the above-mentioned goal. On the other hand, such choices do not distort the non-perturbative content of $P_g(T)$ itself in the whole temperature range. In other words, we need the simulating function $\phi_h(T)$ in order for the gluon plasma pressure and all its derivatives to approach their respective Stefan – Boltzmann limits at high temperatures from below. The simulating functions $f_l(T)$ and $\phi_l(T)$ are needed in order to ensure the continuous character of the gluon plasma pressure across T_c, while all the non-trivial perturbative and non-perturbative physics in the gluon plasma pressure are due to the basic functions $P_{SB}(T)$ and $P_g(T)$, respectively.

9.4 Main thermodynamic quantities

Together with the pressure $P(T)$, the main thermodynamic quantities are the entropy density $s(T)$ and the energy density $\epsilon(T)$. The general formulae which connect them are based on Ref. [184]

$$s(T) = \frac{\partial P(T)}{\partial T},$$
$$\epsilon(T) = T\left(\frac{\partial P(T)}{\partial T}\right) - P(T) = Ts(T) - P(T) \tag{9.4.1}$$

for pure Yang – Mills fields — when the chemical potential is equal to zero. Let us note that in quantum statistics the pressure $P(T)$ is nothing but the thermodynamic potential $\Omega(T)$ apart from the sign, $P(T) = -\Omega(T) > 0$.

Other thermodynamic quantities of interest are the heat capacity $c_V(T)$ and the velocity of sound squared $c_s^2(T)$, which are defined as follows:

$$c_V(T) = \frac{\partial \epsilon(T)}{\partial T} = T\left(\frac{\partial s(T)}{\partial T}\right), \tag{9.4.2}$$

and

$$c_s^2(T) = \frac{\partial P(T)}{\partial \epsilon(T)} = \frac{s(T)}{c_V(T)}, \tag{9.4.3}$$

which are defined through the second derivative of the pressure. The conformity

$$C(T) = \frac{P(T)}{\epsilon(T)} \tag{9.4.4}$$

mimics the behavior of the velocity of sound squared given by equation (9.4.3) but without involving such differentiation.

A thermodynamic quantity of special interest is the thermal expectation value of the trace of the energy momentum tensor. This trace anomaly relation measures the deviation of the difference

$$\epsilon(T) - 3P(T) = T^5 \frac{\partial}{\partial T}\left(\frac{P(T)}{T^4}\right) \tag{9.4.5}$$

from zero at finite temperatures; in the high temperature limit it must vanish. As a consequence it is very sensitive to the non-perturbative contributions to the equation of state.

It also assists in the temperature dependence of the gluon condensate [224, 227], namely

$$\langle G^2 \rangle_T = \langle G^2 \rangle_0 - [\epsilon(T) - 3P(T)], \tag{9.4.6}$$

where $\langle G^2 \rangle_0 \equiv \langle G^2 \rangle_{T=0} = \langle 0 | (1/4) G^a_{\mu\nu} G^a_{\mu\nu} | 0 \rangle = 0.1052 \text{ GeV}^4$ denotes the gluon condensate at zero temperature. Its numerical value has been discussed in Section 6.7.

The so-called enthalpy density is defined as follows:

$$e(T) = T \frac{\partial P(T)}{\partial T} = Ts(T) = \epsilon(T) + P(T). \tag{9.4.7}$$

This sum is of interest and importance, since it appears in the relativistic hydrodynamics equations of motion as in Refs. [189, 220, 228, 229], making them highly non-linear ones. The curve for it is shown in Fig. 9.3.5, since from the definition (9.4.7) it follows that

$$3e(T)/4T^4 = 3Ts(T)/4T^4 = 3s(T)/4T^3. \tag{9.4.8}$$

9.5 The Stefan – Boltzmann limit

The high-temperature behavior of all the thermodynamic quantities is governed by the Stefan – Boltzmann ideal gas limit, when the matter can be described in terms of non-interacting massless particles (gluons). In this $T \to \infty$ limit, these quantities satisfy special relations, namely

$$\frac{3P_{SB}(T)}{T^4} = \frac{\epsilon_{SB}(T)}{T^4} = \frac{3s_{SB}(T)}{4T^3} = \frac{c_{V(SB)}(T)}{4T^3} = \frac{24}{45}\pi^2 \approx 5.2634, \tag{9.5.1}$$

and

$$C_{SB}(T) = c^2_{s(SB)}(T) = \frac{1}{3}, \quad \text{while} \quad T \to \infty, \tag{9.5.2}$$

on account of the previous relations and their definitions in equation (9.4.3) and (9.4.4). The trace anomaly relation (9.4.5) also satisfies the Stefan–Boltzmann limit, namely

$$\epsilon_{SB}(T) - 3P_{SB}(T) = 0, \quad \text{while} \quad T \to \infty, \tag{9.5.3}$$

as it comes out from the relations (9.5.1). The right-hand side of the relations (9.5.1) we call the common/general Stefan–Boltzmann limit/constant.

9.6 Analytical formulae for the gluon plasma thermodynamic quantities

It is instructive to derive analytically all the necessary formulae for the thermodynamic quantities using the gluon plasma pressure (9.3.1). Differentiating it in accordance with the definition (9.4.1), on account of the relation (9.3.2), one obtains

$$s(T) = \frac{\partial P_g(T)}{\partial T} + \Theta\left(\frac{T_c}{T} - 1\right)\frac{\partial L(T)}{\partial T} + \Theta\left(\frac{T}{T_c} - 1\right)\frac{\partial H(T)}{\partial T}, \tag{9.6.1}$$

omitting here and everywhere below the subscript 'GP' in accordance with the remark made in Subsection 9.3. It is easy to see that the entropy density has a jump at T_c,

$$\Delta s(T_c) = [s(T > T_c) - s(T < T_c)]_{T \to T_c} = \left[\frac{\partial H(T)}{\partial T} - \frac{\partial L(T)}{\partial T}\right]_{T=T_c}, \tag{9.6.2}$$

where the difference in the right-hand-side of this equation is positive.

In the same way for the energy density, one obtains

$$\epsilon(T) = T\frac{\partial P_g(T)}{\partial T} - P_g(T) + \Theta\left(\frac{T_c}{T} - 1\right)\left[T\frac{\partial L(T)}{\partial T} - L(T)\right] + \Theta\left(\frac{T}{T_c} - 1\right)\left[T\frac{\partial H(T)}{\partial T} - H(T)\right]. \tag{9.6.3}$$

The size of the discontinuity in the energy density, where the latent heat is denoted by 'LH', is

$$\epsilon_{LH}(T_c) = \Delta\epsilon(T_c) = [\epsilon(T > T_c) - \epsilon(T < T_c)]_{T \to T_c} = T_c\left[\frac{\partial H(T)}{\partial T} - \frac{\partial L(T)}{\partial T}\right]_{T=T_c}, \tag{9.6.4}$$

and thus it is in agreement with the discontinuity in the entropy density, since from equations (9.6.2) and (9.6.4) it follows that the latent heat is $\epsilon_{LH}(T_c) = \Delta\epsilon(T_c) = T_c \Delta s(T_c)$.

The last independent thermodynamic quantity is the heat capacity defined in equation (9.4.2). Differentiating the entropy density (9.6.1), one finally obtains

$$c_V(T) = T\frac{\partial^2 P_g(T)}{\partial T^2} + \Theta\left(\frac{T_c}{T} - 1\right) T\frac{\partial^2 L(T)}{\partial T^2} + \Theta\left(\frac{T}{T_c} - 1\right) T\frac{\partial^2 H(T)}{\partial T^2}$$
$$- \frac{T_c}{T}\delta\left(\frac{T_c}{T} - 1\right)\frac{\partial L(T)}{\partial T} + \frac{T}{T_c}\delta\left(\frac{T}{T_c} - 1\right)\frac{\partial H(T)}{\partial T}. \tag{9.6.5}$$

The important observation is that the heat capacity has a δ-type singularity, an essential discontinuity, at $T = T_c$, so that the velocity of sound squared (9.4.3) at this point is zero, namely

$$c_s^2(T_c) = \frac{s(T_c)}{c_V(T_c)} = 0. \tag{9.6.6}$$

The analytical expression for the velocity of sound squared (9.4.3) can be found with the help of equations (9.4.1) and (9.4.2).

On account of equations (9.4.1) and (9.3.1), the trace anomaly relation (9.4.5) looks like

$$\epsilon(T) - 3P(T) = T\frac{\partial P_g(T)}{\partial T} - 4P_g(T) + \Theta\left(\frac{T_c}{T} - 1\right)\left[T\frac{\partial L(T)}{\partial T} - 4L(T)\right]$$
$$+ \Theta\left(\frac{T}{T_c} - 1\right)\left[T\frac{\partial H(T)}{\partial T} - 4H(T)\right]. \tag{9.6.7}$$

As mentioned above it assists in the evaluation of the temperature dependence of the gluon condensate (9.4.6).

The sum of the pressure (9.3.1) and the energy density (9.4.1), which is nothing but the above-mentioned enthalpy density (9.4.7), is

$$e(T) = T\,\frac{\partial P_g(T)}{\partial T} + \Theta\left(\frac{T_c}{T} - 1\right) T\,\frac{\partial L(T)}{\partial T} + \Theta\left(\frac{T}{T_c} - 1\right) T\,\frac{\partial H(T)}{\partial T}. \tag{9.6.8}$$

Let us point out that the discontinuities which appear in the derivatives of the pressure are not due to the Θ-functions in equation (9.3.1). They are due to the fact that the derivatives of the auxiliary functions $L(T)$ and $H(T)$ are different from each other and they are not zero at T_c, see equations (9.6.2), (9.6.4), and (9.6.5), respectively. The deep reason for these discontinuities is the principal difference between the independent basic functions $P_{SB}(T)$ and $P_g(T)$ from each other, in terms of which the auxiliary functions have finally been found within our approach as follows:

$$\begin{aligned} L(T) &\\ &= \left[0.0282e^{-19.1(\frac{T_c}{T}-1)} + 0.005e^{-3.4(\frac{T_c}{T}-1)}\right] P_{SB}(T) - e^{-0.0001\frac{T_c}{T}} P_g(T), \\ H(T) &\\ &= \left[1 - \frac{0.36}{1 + 0.55\ln\frac{T}{T_c}}\right] P_{SB}(T) - \left[1.2815 + 0.8349\left(\frac{T_c}{T}\right)^3\right] P_g(T), \end{aligned} \tag{9.6.9}$$

as it comes out from the comparison of equations (9.3.1) and (9.3.39), on account of the relations (9.3.40) and (9.3.41). The pure perturbative contribution $P_{SB}(T)$ is given in the relations (9.5.1), while the non-perturbative contribution $P_g(T)$ is given in equation (9.2.1).

9.7 Double-counting in integer powers of α_s problem

Before going to the concrete calculations of all the thermodynamic quantities, it is necessary to resolve the double-counting in integer powers of α_s problem, which has been mentioned in the previous chapter. For our discussion it is sufficient to consider the gluon plasma pressure at $T > T_c$. It can be taken from equation (9.3.39), namely

$$P(T) = [1 - \alpha_s(T)]\, P_{SB}(T) - 0.3P_g(T), \tag{9.7.1}$$

where

$$\alpha_s(T) = \frac{3.04\alpha_s}{1 + 4.64\alpha_s \ln(T/T_c)}, \quad \text{with} \quad \alpha_s = 0.1184, \tag{9.7.2}$$

as it comes out from the expression (9.3.40). Any derivative of the gluon plasma pressure (9.3.39), and hence of pressure (9.7.1), with respect to the temperature requires the differentiation of the running effective charge (9.7.2). We will denote any derivative with respect to T as follows: for example $P'(T) = \partial P(T)/\partial T$ and so on, for convenience. Thus from the previous equations one finally gets

$$s(T) = P'(T) = [1 - \alpha_s(T)]\, s_{SB}(T) + 0.3815\alpha_s^2(T) s_{SB}(T) - 0.3P_g'(T), \tag{9.7.3}$$

since

$$TP_{SB}'(T) = 4P_{SB}(T) = s_{SB}(T)T = \frac{4}{3}\epsilon_{SB}(T) = \frac{1}{3}Tc_{V(SB)}(T), \tag{9.7.4}$$

which are consequences of the relations (9.5.1), while

$$\alpha_s'(T) = -\frac{1.526}{T}\alpha_s^2(T), \tag{9.7.5}$$

so its expansion begins with the α_s^2-order term. In the same way, for the energy density (9.4.1) one finally obtains

$$\epsilon(T) = [1 - \alpha_s(T)]\,\epsilon_{SB}(T) + 0.508 \cdot \alpha_s^2(T)\epsilon_{SB}(T) - 0.3 \cdot [TP_g'(T) - P_g(T)]. \tag{9.7.6}$$

The expansion in integer powers of a small α_s for the running effective charge $\alpha_s(T)$ (9.7.2) contains all powers of α_s, starting from the α_s-order term. Its derivative, which is proportional to $\alpha_s^2(T)$ (9.7.5), also contains all powers of α_s, starting from the α_s^2-order term. This means that starting from the α_s^2-order term, one obtains the double-counting in each order (apart from the α_s-order term) in the expressions for the gluon plasma entropy and energy densities (9.4.1). The terms containing $\alpha_s^2(T)$ explicitly violate the asymptotic freedom approach of the considered thermodynamic quantities to their respective Stefan – Boltzmann limits, which are determined by the first two terms in the expressions (9.7.3) and (9.7.6). Moreover, such kind of terms are present in the gluon pressure $P_g(T)$ (8.4.1) as well, namely $P_{PT}'(\Delta_R^2; T)$, and $\tilde{P}_{PT}(T)$. Their expansions (8.2.30) and (8.2.29) in integer powers of α_s just begin with the α_s^2-order terms. Thus they also contribute to the double-counting problem, making it even more complicated.

In this connection, it is instructive to analyze the trace anomaly relation (9.4.5) which is the intrinsically non-perturbative quantity and thus has to be free of the perturbative contributions, contaminations and, at the same time, it has to be calculated with the help of the confining effective charge (7.3.2). Combining expressions (9.7.1) and (9.7.6), one obtains

$$\epsilon(T) - 3P(T) = 1.524 \cdot \alpha_s^2(T)P_{SB}(T) - 0.3 \cdot \left[TP_g'(T) - 4P_g(T)\right], \tag{9.7.7}$$

because of the relations (9.7.4). Thus it is not free of the perturbative contaminations due to the first term and the terms mentioned above, which are present in the gluon pressure $P_g(T)$. Moreover, in the presence of the first term in equation (9.7.7) it is legitimate to neglect the non-perturbative contributions in the limit of high temperatures, but then it violates the Stefan – Boltzmann relation (9.5.3) in this limit. Now we are in position to explicitly formulate the following proposal.

Prescription: "In all the thermodynamic quantities the terms whose expansions in integer powers of a small α_s begin with the α_s^2- and higher-order terms should be omitted in the whole temperature range".

This resolves simultaneously at least five problems:

(i) The above-mentioned double-counting in integer powers of α_s.
(ii) It restores the asymptotic freedom approach of all the thermodynamic quantities to their Stefan – Boltzmann limits, thus, they are defined now at the same (first) order of $\alpha_s(T)$.
(iii) It automatically makes the trace anomaly relation and all other intrinsically nonperturbative quantities free of the perturbative contributions.
(iv) It is in agreement with lattice calculations of the trace anomaly at high temperatures.
(v) It makes the gluon pressure $P_g(T)$ a well defined object.

Points (iii) and (v) require a more detailed explanation. To omit the perturbative contributions in the right-hand side of equation (9.7.7) means to subtract them from its left-hand side, and thus to define the intrinsically non-perturbative quantity mentioned above. So one obtains

$$\begin{aligned}[\epsilon(T) - \ 3P(T)]_{INP} \\ &= [\epsilon(T) - 3P(T)] - 1.526\alpha_s^2(T)P_{SB}(T) + 0.3\left[TP_g'(T) - 4P_g(T)\right]_{PT} \\ &= -0.3 \cdot \left[TP_g'(T) - 4P_g(T)\right], \qquad (9.7.8)\end{aligned}$$

where its right-hand side is free of the terms $P'_{PT}(\Delta_R^2;T)$ and $\tilde{P}_{PT}(T)$, thus, they have been included into the $0.3\left[TP_g'(T) - 4P_g(T)\right]_{PT}$ term. Evidently, this is equivalent to simply omit all the perturbative terms in the right-hand side of the initial equation (9.7.7). At the same time, this means that the gluon pressure becomes

$$P_g(T) = P_{NP}(T) + P_{PT}\left(\Delta_R^2;T\right), \qquad (9.7.9)$$

where each term is known: $P_{NP}(T)$ has already been exactly calculated, while $P_{PT}\left(\Delta_R^2;T\right)$ is given by the expansion (8.2.13), and its first contribution denoted as $P_{PT}^s(T)$ in equation (9.2.1) has also been already calculated. The remanning term $P_{PT}(\Delta_R^2;T)$ produces the α_s- and higher-order corrections to its corresponding $\alpha_s^0 = 1$ term which is nothing but $P_{NP}(T)$, more precisely $P_{YM}(T)$. In this connection let us note that there cannot be any perturbative corrections to the bag constant itself $B_{YM}(T)$, since its skeleton part (7.2.4) has been defined from the very beginning by the subtraction of all the types of the perturbative contributions. In other words, $P_{PT}(\Delta_R^2;T)$, conventionally denoted as the perturbative term, is in fact the non-perturbative contribution, which is only suppressed by integer powers of a small α_s. That is why it should be kept alive in the intrinsically non-perturbative trace anomaly relation (9.7.8). Its expansion (8.2.13) is convergent, and thus it is uniquely defined — there is no ambiguity in our formalism after adopting our proposal/prescription. This means that the gluon pressure (9.7.9) is also exactly known.

In the previous Chapter we developed the formalism which made it possible to calculate the perturbative contributions to the equation of state in terms of the convergent series in integer powers of a small α_s. In this way we have resolved the long-term problem of a non-analytical dependence on α_s in the initial thermal perturbation theory QCD whose expansion turned out to be divergent, though each term up to $\alpha_s^3 \ln \alpha_s$-order has been calculated correctly, as underlined in the previous Chapter. Resolving this problem, we have encountered the problem of the unnecessary terms $P'_{PT}(\Delta_R^2;T)$ and $\tilde{P}_{PT}(T)$ whose convergent expansions in integer powers of α_s contribute to the above-discussed double-counting problem. Apparently, this is the price we have to pay for the solution of a non-analytical dependence on the coupling constant in the thermal perturbation theory expansion of the equation of state. However, the prescription formulated above resolves this and many other problems within our approach to QCD at non-zero temperature.

9.8 Numerical results and discussion

The gluon plasma pressure

We present the gluon plasma pressure (9.3.39) as well as the lattice pressure given in Ref. [221] and in Fig. 9.3.4. The close shapes of both curves are remarkable, which is surprising since we have matched our simulations procedure with the lattice data at high temperature only, starting approximately from $3T_c$. We can now predict the value of the gluon plasma pressure, and hence of all other thermodynamic quantities, in the region of very low temperatures, where lattice uncertainties still remain very big. This problem does not exist within our analytical approach to the equation of state for the gluon plasma. In other words, for any given value of temperature from this region we are able to give a concrete number for any thermodynamic quantity, see figures below. However, the most important problem of lattice simulations is solved. Now we know what is the physics behind all the lattice curves and their numbers.

One of the interesting features of all the lattice simulations — see, for example Refs. [221, 224]) is a rather slow approach to the common Stefan–Boltzmann limits (9.5.1) at high temperatures of all the independent thermodynamic quantities. Within our formalism the regime at high temperatures is controlled by the running coupling constant $\alpha_s(T)$ (9.3.40), which depends on T only logarithmically. As emphasized above, all the independent thermodynamic quantities correctly calculated within non-perturbative analytic equation of state (9.3.39) should approach the ideal gas limit in the asymptotic freedom way, slowly and from below. It is instructive to explain this issue, for further purpose as well, in more detail. The pressure (9.3.39) above T_c is

$$P(T) = (1 - \alpha_s(T))\, P_{SB}(T) - 0.3 \cdot P_g(T), \quad \text{while} \quad T > T_c, \qquad (9.8.1)$$

where we omit the next-to-leading term $\sim T^{-3}$ in equation (9.3.41), since this plays no role in the present discussion, and replace 1.2815 by 1.3, for simplicity. In dimensionless units (9.3.19) it looks like

$$\frac{3P(T)}{T^4} = 5.2634 \cdot [1 - \alpha_s(T)] - 0.3 \cdot \frac{3P_g(T)}{T^4}, \quad \text{while} \quad T > T_c, \qquad (9.8.2)$$

because of equation (9.5.1). From our numerical results it follows that both terms in this equation are approximately of the same order of magnitude in the moderately high temperatures interval up to $(3-4)T_c$, thus the negative contribution of the second term still plays a significant role. Moreover, T_c is fixed by $P_g(T)$ itself and the shape of the gluon plasma pressure in Fig. 9.3.4 just above T_c is determined by $P_g(T)$ as well. It is worth emphasizing once more that we have reproduced the lattice pressure at T_c very well, adjusting to lattice data far away from T_c — compare our value (9.3.34) with the lattice one (9.3.35). There is no doubt that the gluon pressure $P_g(T)$ correctly reproduces the non-perturbative structure of the full gluon plasma pressure $P_{GP}(T)$ in the whole temperature range. The addition of the positive perturbative term — which varies very slowly in the whole temperature range above T_c — to $P_g(T)$ in equation (9.8.2) cannot provide such a sharp change in the behavior of the gluon plasma pressure just above the critical temperature, T_c.

On the contrary, in the limit of high temperatures the first perturbative term will become dominant, but this is not the whole story yet. Indeed, from our previous Chapter the corresponding asymptotic of the gluon pressure $P_g(T)$ (8.6.35) is as follows:

$$P_g(T) \sim B_2 \alpha_s \Delta_R^2 T^2 + [B_3 \Delta_R^3 + M^3] T, \quad T \to \infty, \qquad (9.8.3)$$

where the first leading term, which analytically depends on the mass gap Δ_R^2, comes from the perturbative part of the gluon pressure, more precisely from the non-perturbative part which is α_s-order suppressed. The second term's dependence on the mass gap is not analytical, since $\left(\Delta_R^2\right)^{3/2} = \Delta_R^3$, comes from both parts of the gluon pressure. Note that the overall numerical factors B_2 and B_3 are not important here. M^3 denotes the terms of the dimensions of the GeV^3, which depend analytically on the mass gap Δ_R^2. Thus in dimensionless units, as $T \to \infty$, one obtains

$$\frac{3P(T)}{T^4} \sim [1 - \alpha_s(T)] \cdot 5.2634 - 0.9 \cdot \left[\bar{B}_2 \alpha_s \left(\frac{T_c}{T}\right)^2 + \bar{B}_3 \left(\frac{T_c}{T}\right)^3 \right], \quad (9.8.4)$$

where $\bar{B}_2 = B_2(\Delta_R/T_c)^2$ and $\bar{B}_3 = B_3(\Delta_R/T_c)^3 + (M/T_c)^3$. Let us remind that the mass gap term $\Delta_R^2 T^2$ explicitly shown in equation (9.2.2) is canceled within $P_{NP}(T)$ at high temperatures, and so only its α_s-suppressed counterpart explicitly shown in equation (9.8.4) survives in this limit, it comes from $P_{PT}^s(T)$. However, as it follows from our numerical results at moderately high temperatures the negative contribution from the second term in equation (9.8.2) still remains substantial. And only in the limit of very high temperatures the power-type corrections of the second term in equation (9.8.4) become small in comparison with the contribution of the first term, which is of a logarithmical dependence on T. So the combination of these two effects in the whole temperature range just explains the above-mentioned slow approach of the gluon plasma pressure given by (9.3.39) and its derivatives to the common Stefan – Boltzmann limit, clearly seen in Figs. 9.3.5 and 9.8.1.

Energy and entropy densities

The gluon plasma entropy and energy densities (9.4.1) are shown in Fig. 9.3.5. The size of the discontinuity in the energy density, the so-called latent heat ('LH'), is

$$\epsilon_{LH} = 1.414 \quad (9.8.5)$$

in dimensionless units — see Section 9.6 and Appendix 9.A for its definition and analytical/numerical evaluation, respectively. This means that the first-order phase transition in the gluon plasma is analytically confirmed for the first time, in complete agreement with thermal $SU(3)$ Yang – Mills lattice simulations [221, 224]. The reason of such sharp changes at T_c in the derivatives of the gluon plasma pressure is that its exponential rise below T_c is changing to the polynomial rise above T_c in order to reach finally the Stefan – Boltzmann limit. The value (9.8.5) is in fair agreement with recent lattice ones: in Refs. [221] $\epsilon_{LH} = 1.4$, in [230] $\epsilon_{LH} = 1.413$ and [231] $\epsilon_{LH} = 1.44$, and differs from the old one, which is $\epsilon_{LH} \approx 2$ — see Ref. [224]. This agreement is not a trivial thing, since, as mentioned above, we have adjusted our analytical numerical simulations with those of lattice ones in Ref. [221] only for the pressure and far away from T_c. First of all, the energy and entropy densities are the independent thermodynamic variables. Secondly, the lattice results heavily depend on how the continuum limit is to be taken from Ref. [231] and on other details of lattice simulations, while our approach is free of such kinds of problems. And thirdly, it is important to remind that in all lattice simulations the β-function is completely different from ours (as written in Appendix 9.B), it is always negative as it is required by confinement, Section 4.4. The slow approach of the energy and entropy densities to their common Stefan – Boltzmann limit plotted on Fig. 9.3.5 has already been explained in the previous Subsection.

Heat capacity

The last independent thermodynamic quantity the heat capacity, defined in equation (9.4.2), is shown in Fig. 9.8.1. It is always a smoothly growing function of T, both below and above T_c, while at T_c it has a δ-type singularity (an essential discontinuity) due to the expression (9.6.5). It very slowly approaches the common Stefan – Boltzmann limit (9.5.1) at high temperatures.

Conformality, conformity and the velocity of sound squared

The gluon plasma pressure versus the gluon plasma energy density, $P(\epsilon)$, is present in Fig. 9.8.2. The size of the latent heat and a rather rapid approach to conformality are clearly seen. We distinguish between conformality here and conformity defined in equation (9.4.4), though numerically in the limit of high temperatures they are the same. Conformity itself is shown in Fig. 9.8.3. It has a finite negative jump at T_c because of a jump in the energy density at this point, and it rather rapidly approaches its Stefan – Boltzmann limit (9.5.2) at high temperatures. However, its most interesting feature is a non-trivial dependence on T below T_c, which has been fixed explicitly in $SU(3)$ gluon plasma for the first time. On the one

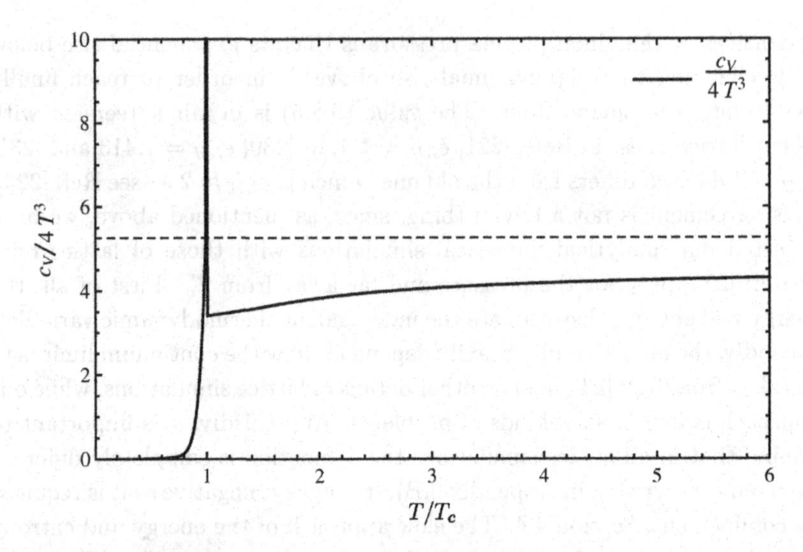

Fig. 9.8.1 The heat capacity (9.4.2) is shown as a function of T/T_c. It has a δ-type singularity — as an essential discontinuity — at T_c. It very slowly approaches the general Stefan – Boltzmann constant (9.5.1) (*horizontal dashed line*) at high temperatures.

hand, it can be due to the fact that conformity is the ratio of the independent thermodynamic quantities — for it the exponential suppression at low temperature is not obligatory. On the other hand, the protuberance in the region $(0.2 - 0.4)T_c$ is due to the complicated non-perturbative structure of the gluon pressure $P_g(T)$ and its derivatives. It dominates the structure of the gluon plasma pressure below T_c within the analytical approach — the Stefan – Boltzmann-type terms cannot penetrate so deeply into the low-temperature region. In principle, the shape of the curve below T_c (see Figs. 9.8.4 and 9.8.3) may be changed (or not) by the more detailed approximation of $P_{GP}(T)$ by $P_g(T)$ in this region. However, numbers are rather small, and the problem (the existence of the protuberance) seems not to be so important from the numerical point of view.

In close connection with the previous quantities is the velocity of sound squared (9.4.3) shown in Fig. 9.8.4. Below T_c it behaves very similarly to conformity Fig. 9.8.3, since the latter one mimics its properties. The principal difference from conformity is that at T_c it is zero because of the heat capacity having the above-mentioned δ-type singularity at this point. It rather rapidly approaches its Stefan – Boltzmann limit (9.5.2) at high temperatures.

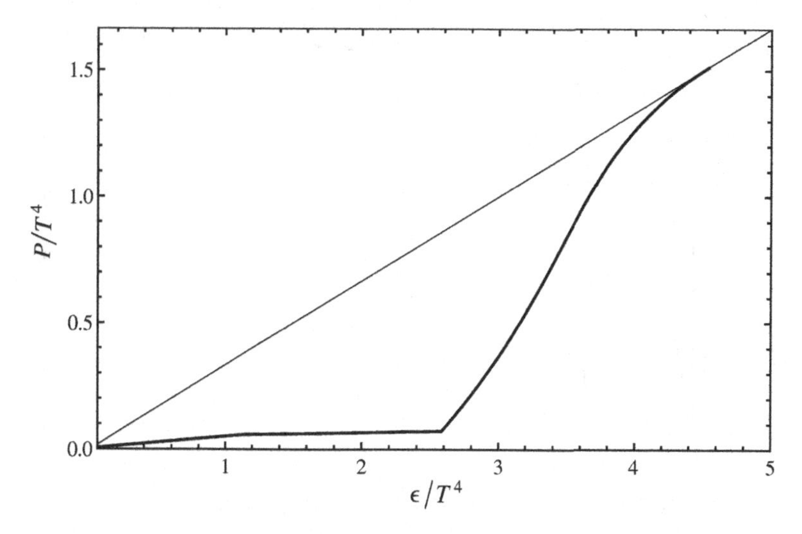

Fig. 9.8.2 The equation of state, $P(\epsilon)$ (*solid line*) and rather rapid approach to conformality = 1/3 (*diagonal thin line*).

Trace anomaly relation

The trace anomaly, defined in equation (9.4.5), is especially sensitive to the non-perturbative effects. In fact, it is the intrinsically non-perturbative quantity within our approach as it can be seen in Section 9.7. It is shown in Figs. 9.8.5, 9.8.6, and 9.8.7. Its three most interesting features are:

(i) The rapid rise of the peak due to the LH in the energy density plotted on Fig. 9.3.5 is exactly placed at T_c, and it is about 2.5, see Fig. 9.8.5. In all lattice calculations it peaks at about $1.1T_c$ as in Refs. [219–221, 224, 231], and it is about 2.6. The wrong position of the trace anomaly lattice peak can be due to an ultraviolet cutoff, the finite volume effects, etc. In this connection let us remind that in lattice simulations at any temperature the aim is to go finally to the continuum physical limit, namely lattice spacing goes to zero and then the infinite volume limit should be taken. These are nothing but the removal of the ultraviolet and infrared cutoffs which is part of the renormalization procedure as it is written in Refs. [77, 146]. This means that lattice results for any non-perturbative quantity, in particular the trace anomaly will always be contaminated by the additional contributions of perturbative origin. Of course, they may

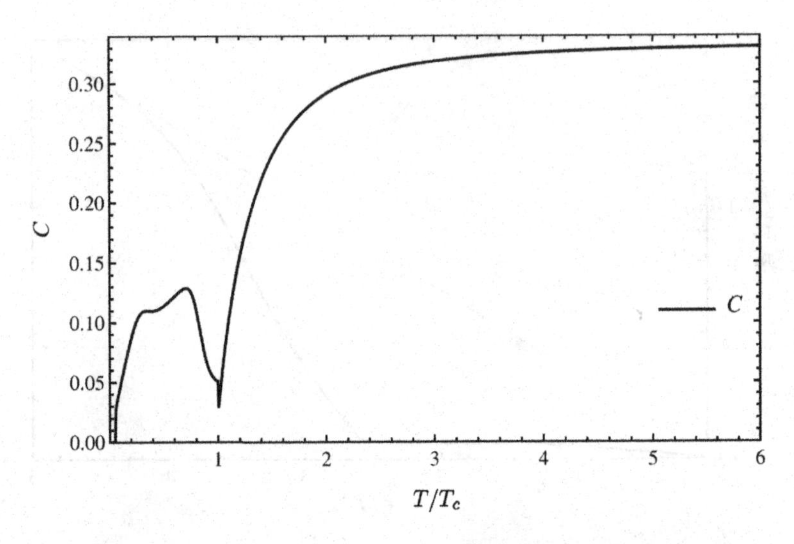

Fig. 9.8.3 Conformity given by (9.4.4) is shown as a function of T/T_c. It is zero at $T = 0$, and has a finite negative jump at T_c due to a jump in the energy density at this point. It shows a non-trivial dependence on T below T_c, and rather rapidly approaches the Stefan–Boltzmann limit (9.5.2) at high temperatures.

produce some changes in the non-perturbative structure of the lattice trace anomaly, for example shifting effectively the position of its peak. However, we think that this problem may signify deep dynamical, non-perturbative properties of the trace anomaly, in general, and near T_c, in particular.

(ii) The second interesting feature is that even just above T_c and up to rather high temperatures the non-perturbative effects due to the mass gap are still important in the trace anomaly. This is clearly seen in Figs. 9.8.6 and 9.8.7, which demonstrate complicated dependence of the trace anomaly on the mass gap and the temperature in this interval. Both curves shown in these figures approach the corresponding constants at very high temperatures only. Indeed, the trace anomaly equation (9.7.8) is

$$\epsilon(T) - 3P(T) = -0.3 \cdot \left[TP_g'(T) - 4P_g(T)\right], \tag{9.8.6}$$

where the subscript 'INP' has been omitted, for simplicity. In accordance with this equation the above-described nontrivial structure of the trace anomaly above T_c is determined by the gluon pressure $P_g(T)$, given now in equation (9.7.9), and its derivative. In order to calculate Fig. 9.8.6 we repeat the Pisarski trick written in Ref. [219],

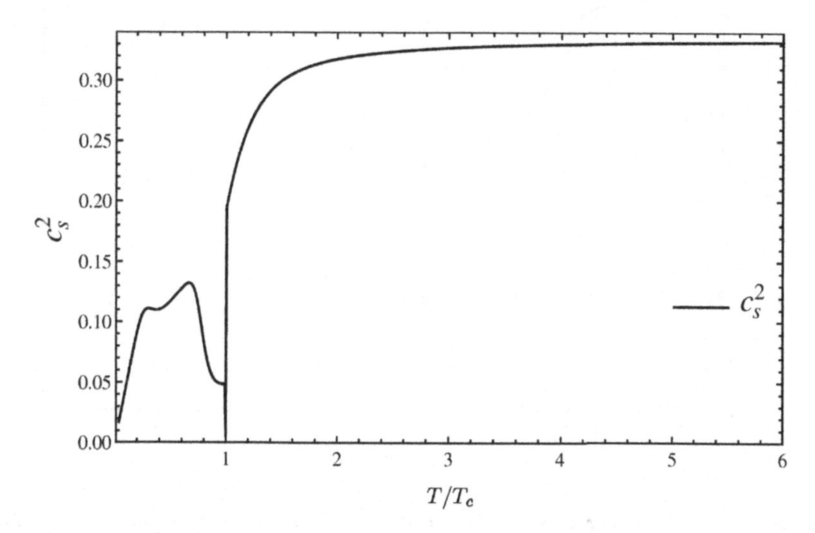

Fig. 9.8.4 The velocity of sound squared (9.4.3) is shown as a function of T/T_c. It shows a non-trivial dependence on T below T_c, while at $T = 0$ and at $T = T_c$ it is zero. It rather rapidly approaches the Stefan – Boltzmann limit (9.5.2) at high temperatures.

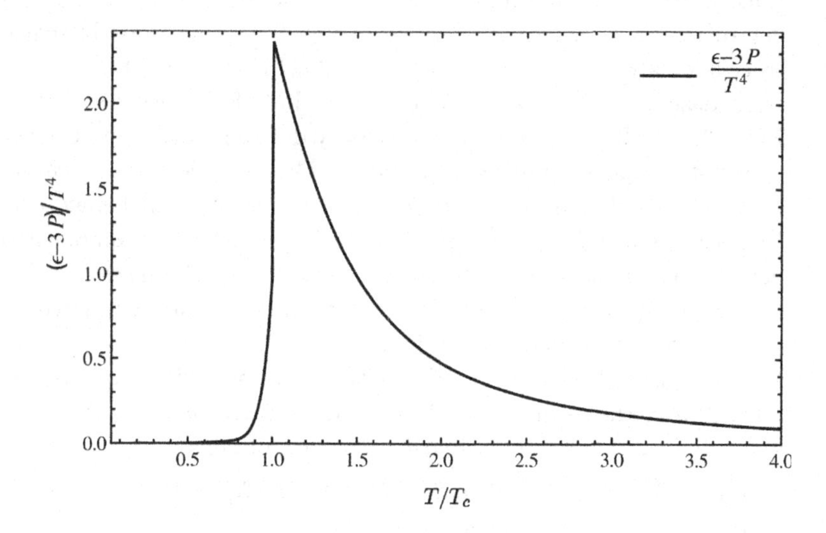

Fig. 9.8.5 The trace anomaly (9.4.5) properly scaled is shown as a function of T/T_c.

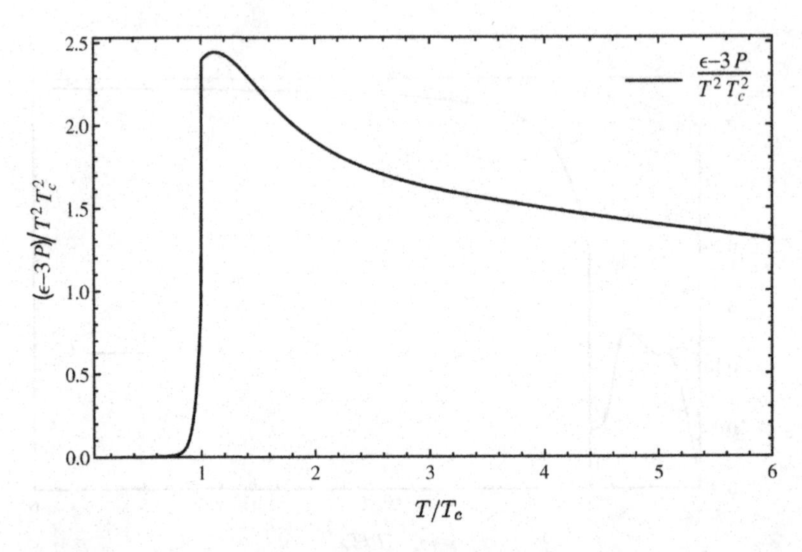

Fig. 9.8.6 The trace anomaly (9.4.5) scaled by $T^2T_c^2$ is shown as a function of T/T_c. It demonstrates a non-trivial structure just above T_c, and it slowly approaches the corresponding constant in the limit of very high temperatures only, see equation (9.8.9).

thus, we multiply the properly scaled trace anomaly by $T^2T_c^{-2}$ in order to remain dimensionless, while in Fig. 9.8.7 the properly scaled trace anomaly is shown as a function of $(T_c/T)^2$. In Fig. 9.8.7 the constant dashed line reproduces nothing else but the lattice result from Refs. [221, 222] — apart from the slightly different height and the position of the peak, as discussed above. The deviation of our curves from the lattice line in Fig. 9.8.7 and from the constant behavior in Fig. 9.8.6 above T_c is obvious. It is mainly due to the complicated dependence of the gluon pressure $P_g(T)$ (9.7.9) on the mass gap and the temperature in this region, where it cannot be approximated by some simple power-type expression.

(iii) However, this is possible to do in the limit of high temperatures. Substituting the asymptotic (9.8.3) and its derivative into the trace anomaly equation (9.8.6), while $T \to \infty$, one obtains

$$\epsilon(T) - 3P(T) \sim 0.6 \cdot B_2\alpha_s\Delta_R^2T^2 + 0.9 \cdot \left[B_3\Delta_R^3 + M^3\right]T, \quad (9.8.7)$$

and in dimensionless units it becomes

$$\frac{\epsilon(T) - 3P(T)}{T^4} \sim 0.6 \cdot \bar{B}_2\alpha_s\left(\frac{T_c}{T}\right)^2 + 0.9 \cdot \bar{B}_3\left(\frac{T_c}{T}\right)^3, \quad \text{while} \quad T \to \infty, \quad (9.8.8)$$

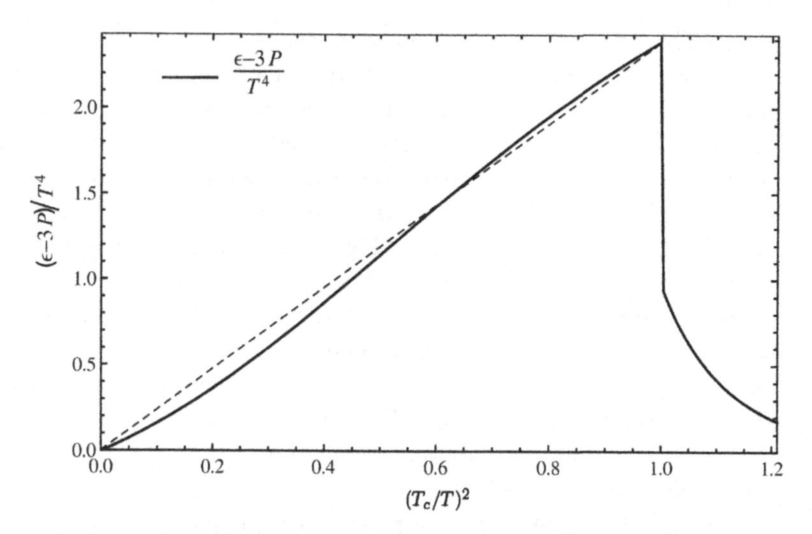

Fig. 9.8.7 The trace anomaly (9.4.5) properly scaled is shown as a function of $(T_c/T)^2$ (*solid line*). It approaches the corresponding constant (*dashed line*) in the limit of very high temperatures only.

and hence

$$\frac{\epsilon(T) - 3P(T)}{T^2 T_c^2} \sim 0.6 \cdot \bar{B}_2 \alpha_s + 0.9 \cdot \bar{B}_3 \left(\frac{T_c}{T}\right), \quad \text{while} \quad T \to \infty, \tag{9.8.9}$$

where $\bar{B}_2 = B_2(\Delta_R/T_c)^2$ and $\bar{B}_3 = B_3(\Delta_R/T_c)^3 + (M/T_c)^3$, while for the coefficients B_2, B_3 and the quantity M, see text after equation (9.8.4).

In Ref. [222] the lattice results for the trace anomaly have been analyzed in detail. It has clearly been shown that lattice results exclude odd powers of (T_c/T) and are taken into account only for its even powers above T_c. So the last term in the expansion (9.8.8) cannot be taken into account by lattice simulations, only the first term in this equation can. Our explanation is as follows: the first term comes from the non-perturbative part denoted as $P^s_{PT}(\Delta^2_R; T)$, and see discussion in Section 9.7 in the gluon pressure $P_g(T)$ (9.7.9), in which the analytical dependence on the mass gap is suppressed by the α_s-order. Apparently, that is a reason why lattice simulations may take it into account. The second term in equation (9.8.9) is completely of the non-perturbative dynamical origin, non-analytically depending on the mass gap, and that is why it is out of reach by lattice simulations. The existence of such kinds of terms is a clear evidence of the complicated dynamical structure of trace anomaly not only at high temperature but near T_c as well. However, the first term is dominant in the $T \to \infty$ limit. So the trace anomaly (9.8.9) falls off in this limit exactly as $1/T^2$ in complete agreement with lattice simulations pointed out above. And this is its third characteristic feature mentioned above. Based on the plot of Fig. 9.8.7 it clearly follows that the constant regime begins at rather high temperatures, since the constant term is α_s-order suppressed, see equation (9.8.9). That is the reason why the approach to the constant regime is so slow in Fig. 9.8.6. In turn, this means that the fall off of the trace anomaly just above T_c, shown in Fig. 9.8.5, is not $\sim 1/T^2$, but has much more complicated character, as emphasized above. The fall off $\sim 1/T^2$ starts at rather high temperature, indeed.

Due to the complicated non-perturbative, dynamical structure of the trace anomaly in the whole temperature range, which follows from our non-perturbative analytical approach, it is difficult indeed to believe that its simple power-type behavior starts just from T_c, and especially that it is only of T^2-type. However, the absence of the contribution proportional to odd power of (T_c/T) in lattice results may effectively shift the position of the lattice peak from T_c to $1.1T_c$. Such a possibility is clearly seen in Fig. 9.8.6, where the peak is exactly placed at $1.1T_c$. In any case, the wrong position of the lattice peak is an evidence of the complicated dynamical structure of the trace anomaly near T_c, and it cannot be accounted for by lattice simulations in all details. This is our opinion on this lattice problem, which is subject to discussion, of course. From the general point of view, we attribute all these differences to the fact that our confining effective charge, and hence the corresponding β-function, as in Appendix 9.B, is completely different from those used in the above-cited lattice simulations at finite temperature. Since the trace anomaly is especially sensitive to the non-perturbative effects, the role of the confining β-function at finite temperature remains important. We think that in any case the perturbative contamination mentioned above has to be numerically much smaller than the non-perturbative contribution(s) discussed here.

Concluding, a few remarks are in order. Only the trace anomaly defined by equation (9.8.6) agrees with lattice results at high temperatures. Otherwise it would be growing as T^2 function in this limit because of the first term in equation (9.7.7) divided by $T^2T_c^2$. It is the mass gap which violates the conformal symmetry from the very beginning within our approach. Due to the sensitivity of the trace anomaly to non-perturbative effects, it shows explicitly the very non-trivial dynamical structure of the gluon plasma.

Gluon condensate

In close connection with the trace anomaly is the gluon condensate defined by equation (9.4.6) and shown in Fig. 9.8.8. It approaches zero from below, so it gains very small negative values at high temperatures — fixed also by lattice simulations in Ref. [224]. This is due to the fact that the trace anomaly enters equation (9.4.6) with negative sign, and it falls off as $1/T^2$ only in the limit of high temperatures. The gluon condensate at zero temperature has been defined and calculated by the subtraction of all types of the perturbative contributions/contaminations in Section 6.7. Here the temperature-dependent gluon condensate has been calculated with the help of the trace anomaly relation, which is also free of the perturbative contributions. Apparently, it makes sense to consider it as one of the important non-perturbative ingredients in the dynamical structure of the gluon plasma. The gluon plasma pressure has been calculated by dropping the $\beta = 1/T$-independent terms [184]. In Section 9.7, we have explained that dropping terms in one side of the equation means to subtract them from its other side. So adopting this prescription, the regularized, 'R' gluon condensate simply becomes $\langle G^2\rangle_T^R = \langle G^2\rangle_T - \langle G^2\rangle_0 = -\left[\epsilon(T) - 3P(T)\right]$. Properly scaled it can be seen as in Fig. 9.8.5 but with an overall negative sign.

9.9 The dynamical structure of $SU(3)$ gluon plasma

A rich dynamical structure of the gluon plasma emerges in the framework of our approach. The gluon plasma pressure (9.3.39) describes the three different types of massive gluonic excitations. They are: ω' and $\bar{\omega}$ with the effective masses $m'_{eff} = 1.17$ GeV and $\bar{m}_{eff} = 0.585$ GeV, respectively. The third one is again $\bar{\omega}$ whose propagation, however, is suppressed by the α_s-order. We have denoted it as $\alpha_s \cdot \bar{\omega}$. As mentioned above, we can treat it as a new massive excitation, but with the same effective mass $\bar{m}_{eff} = 0.585$ GeV. Both effective masses are due to the mass gap Δ^2 — they have not been introduced by hand. The mass gap itself is dynamically

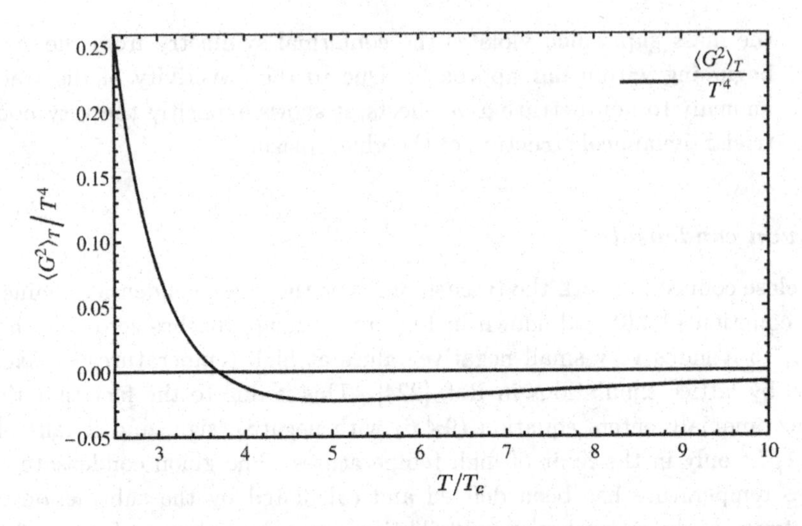

Fig. 9.8.8 The properly scaled gluon condensate at non-zero temperature (9.4.6) is shown as a function of T/T_c. It approaches zero from below at high temperatures.

generated by the nonlinear interaction of massless gluon modes (Part I). The first effective mass m'_{eff} is comparable with the masses of scalar glueballs. The effective mass $\bar{m}_{eff}$ might be identified with an effective gluon mass of about (500 – 800) MeV, which arises in different quasi-particle models [232, 233]. We also have the two different massless gluonic excitations ω, conventionally denoted as ω_1 and ω_2. The former describes the massless gluons, propagating in accordance with the integral $P_1(T)$ in equation (9.2.3). The latter one describes the massless gluons, propagating in accordance with the integral $P_2(T)$ in equation (9.2.5). Let us remind that the integral (9.2.3) should be multiplied by $(6/\pi^2)\Delta_R^2$, and all other integrals (9.2.5) are to be multiplied by $(16/\pi^2)T$, when one speaks about different non-perturbative contributions to the pressure (9.2.2). The propagation of the massive gluonic excitation $\alpha_s \cdot \bar{\omega}$ has to be understood in the same way. The corresponding integral should be multiplied by an overall numerical factor, see equation (9.2.9).

In the low-temperature $T \to 0$ limit all these excitations are exponentially suppressed, since the low-temperature asymptotic of the gluon plasma pressure is determined by the gluon pressure $P_g(T)$, see equations (9.3.6). The dependence on the mass gap and the temperature of the terms, describing their propagation at finite temperatures, is rather complicated. In the high-temperature $T \to \infty$ limit the suppression is only power-type, if the gluon plasma pressure is scaled by $T^4/3$. Let us remind that in both limits a non-analytical dependence on the mass gap occurs, but it is not an expansion parameter like α_s. So in the different temperature intervals they propagate in different ways.

It is important to understand that the above-mentioned effective masses are not the pole masses which appear in the corresponding propagators (see, for example Refs. [184, 234]). This means that we cannot assign to the corresponding massive excitations a meaning of physical particles. They have to be treated rather as quasi-particles. They appear through the corresponding gluon mean numbers, something like the quark chemical potentials. Indeed, from equation (9.2.8) one gets

$$N'_g \equiv N_g(\beta, \omega') = \frac{1}{e^{\beta\omega'} - 1} = \frac{1}{e^{\beta\sqrt{\omega^2 + m'^2_{eff}}} - 1} = \frac{1}{e^{\beta(\omega - \mu'_g)} - 1} \tag{9.9.1}$$

where we introduce the fictitious gluon chemical potential μ'_g. It has to satisfy the following equation $\mu'^2_g - 2\omega\mu'_g - m'^2_{eff} = 0$, which has the two independent solutions: $\mu'_g = \omega \pm \omega' = \omega \pm \sqrt{\omega^2 + m'^2_{eff}}$, leading, nevertheless, to the same effective mass m'^2_{eff}, but only the solution $\mu'_g = \omega - \omega'$ is compatible with equation (9.9.1). By making the replacement $\omega' \to \bar{\omega}$ in equation (9.9.1), we can treat the massive gluonic excitation $\bar{\omega}$ in the same way as ω'. Again, one has the two independent solutions for the fictitious gluon chemical potential $\bar{\mu}_g = \omega \pm \bar{\omega} = \omega \pm \sqrt{\omega^2 + \bar{m}^2_{eff}}$, leading, nevertheless, to the same effective mass $\bar{m}^2_{eff}$, but only the solution $\bar{\mu}_g = \omega - \bar{\omega}$ will be compatible with the corresponding equation (9.9.1). In the excitation $\alpha_s \cdot \bar{\omega}$ the effective mass $\bar{m}^2_{eff}$ appears not only through the corresponding gluon mean number, but in a more complicated way, see equation (9.2.9). For convenience, we denote its chemical potential as $\alpha_s \cdot \bar{\mu}_g$. All three gluon chemical potentials μ'_g, $\bar{\mu}_g$, and $\alpha_s \cdot \bar{\mu}_g$ differ from each other by the corresponding effective masses and by the ranges for ω — see integrals $P_3(T)$, $P_4(T)$ in equation (9.2.5) and integral (9.2.9). Evidently, the corresponding gluon chemical potentials for the two massless excitations ω_1 and ω_2 are zero, thus, $\mu_1 = \mu_2 = 0$ with the same range for ω, see integral (9.2.3) and the first of the integrals (9.2.5).

In principle, we can interpret our effective excitations as the gluon 'flavors', but better to use the term 'species'. So we have the five different gluonic species, which are present in the gluon plasma. Contrary to the quark flavors, all our species are of the non-perturbative dynamical origin, since in the $\Delta^2 = 0$ limit they disappear from the gluon plasma spectrum

— the dependence of the massive species μ'_g, $\bar{\mu}_g$ and $\alpha_s \cdot \bar{\mu}_g$ on the range for ω only confirms this. In other words, it is better to treat our massive excitations/species as some kind of quasi-particles, created by the self-interaction of massless gluon modes at non-zero temperature, thus, consisting of the gluon matter only. That these masses are very close to scalar glueballs and Debye screening masses may or may not be a coincidence, but there are no other massive excitations in the gluon plasma within our approach. Let us remind that their values are in good agreement with established thermal perturbative QCD results [184, 232, 233].

At present, nobody can definitely answer the question why some gluons acquire a mass and some others not. At finite temperatures some gluon fields may intensively interact with each other, leading thus to the formation of stable gluon field configurations, the so-called stationary states with the minimum of energy. A possible existence of such kinds of states of purely transversal virtual gluon field configurations in the QCD vacuum at zero temperature has been discussed in detail in Section 6.7. At non-zero temperature the above described stationary states might be also formed/created, and effectively they can be considered as the massive gluonic excitations. If the self-interaction of massless gluon modes is very intense and an effective mass is big enough then such a stable configuration can be treated as a glueball. If the self-interaction is not intense then an effective mass is not so big. Such a configuration may be considered as a massive gluon. If the self-interaction can be neglected, then the gluons remain massless. In any case, the different types of the massive and massless gluonic excitations of the dynamical origin will necessarily appear at non-zero temperature [184, 199, 215, 216, 225, 232, 233, 235].

The gluon plasma pressure (9.3.39) assumes the presence of the two different perturbative massless excitations, propagating above T_c. The first one is described by the Stefan–Boltzmann term $P_{SB}(T)$ itself, while the second one by its asymptotic freedom correction denoted as $\alpha_s(T) \cdot P_{SB}(T)$. We necessarily also have the two types of the excitations contributing to the gluon plasma pressure below T_c, conventionally called the Stefan–Boltzmann-type terms. On the one hand, they can be of non-perturbative origin, but not explicitly depending on the mass gap, because of the exponential suppression and rise in the $T \to 0$ and $T \to T_c$ limits, respectively. The behavior of the pure perturbative contributions as functions of T is completely different, in general, and in the $T \to T_c$ limit, in particular. If so then they describe the non-perturbative contributions which cannot be taken into account by the gluon pressure $P_g(T)$. On the other hand, they can be interpreted as massless gluonic excitations because of the explicit presence of the Stefan – Boltzmann pressure $P_{SB}(T)$ in these terms, but nevertheless, the propagations of these terms are exponentially suppressed in this region. There is no possibility to clearly distinguish between these two interpretations, since we cannot fix the dynamical or other context of the parameters A_i and α_i. Within our approach we were able to fix only their numerical values (9.3.38), which was fully sufficient to establish the behavior of the gluon plasma pressure below T_c and at T_c with a rather good accuracy — up to second digit after point.

The $P_g(T)$ describes the changes in the regime of the gluon plasma pressure's behavior near T_c, namely the exponential rise transforms to the polynomial one, providing a continuous transition of the gluon plasma pressure across T_c with the help of the Stefan – Boltzmann-type terms. It is necessary to admit that the Stefan – Boltzmann part above T_c is empirically restored to the gluon plasma pressure (9.3.39), but its behavior at high temperatures agrees with lattice simulations, as described above. All the non-perturbative massive and massless gluonic excitations/species have not been introduced by hand; on the contrary, they are of the dynamical origin due to the confining effective charge (7.3.2). They are described and accounted for by the gluon pressure $P_g(T)$ within the non-perturbative analytic approach to thermal QCD.

The exponential rise of all the independent thermodynamic quantities in the transition region $(0.8 - 1)T_c$ clearly seen in Figs. 9.3.3, 9.3.4, 9.3.5, and 9.8.1 indicates that near T_c a dramatic increase in the number of effective gluonic degrees of freedom will appear — and we know that this is so indeed [235]. The massive excitations/species will begin to rapidly dissolve. This will lead to drastic changes of the structure of the gluon plasma. A change in this number is enough to generate pressure gradients, but not enough to affect the pressure itself. It varies slowly and therefore remains continuous in this region. At the same time, the pressure gradients such as the energy and entropy densities, etc., undergo sharp changes in their behavior, having different types of discontinuities at T_c. Thus $SU(3)$ gluon plasma has a first-order phase transition within our approach, in complete agreement with the thermal QCD lattice calculations described in Refs. [221, 224]. Of course, not all the massive excitations will be dissolved in the transition region. Some of them will remain above T_c together with other gluonic excitations and effective gluonic degrees of freedom, which may include the above-mentioned different perturbative contributions as well as gluon condensates. This forms a mixed phase around T_c [215, 236].

One can conclude that the non-perturbative physics of the mixed phase, the temperature interval approximately starting from $0.8T_c$ up to $(3-4)T_c$, is well understood within our approach. Moreover, the region of low temperatures, where all the independent thermodynamic variables are exponentially suppressed at $T \to 0$, is also determined by it, which correctly takes into account all the non-perturbative effects in the gluon plasma. In the mixed phase the gluon plasma can be considered as being in the strong coupling regime. Beyond it the non-perturbative effects become small, and the gluon plasma can be considered as being in the weak coupling regime. The structure of the gluon plasma will be mainly determined by the Stefan – Boltzmann relations between all the thermodynamic quantities at very high temperatures only (9.5.1)–(9.5.3), apart from the trace anomaly.

9.10 Conclusions

The effective potential approach for composite operators provides a new general analytic approach to QCD at zero and non-zero temperature and density [156, 61, 215, 235]. It is essentially non-perturbative in origin, but incorporates the thermal perturbation theory in a self-consistent way. The non-perturbative analytical equation of state for pure $SU(3)$ Yang – Mills fields has been derived within the framework of this formalism. In its non-perturbative part there is no dependence on the coupling constant, only a dependence on the mass gap, which is responsible for the large-scale structure of the QCD ground state. Its perturbative part does analytically depend on the fine-structure constant of strong interactions α_s as well. We have developed the analytic thermal perturbation theory in form of the convergent series, which made it possible to calculate the perturbative part of the equation of state termwise in integer powers of a small α_s. Here we have completed our work by proposing and formulating a method to include the massless free gluons contribution into the Yang-Mills equation of state or, equivalently, gluon plasma equation of state in a self-consistent way. This concludes the derivation of the non-perturbative analytical full gluon plasma equation of state, describing the dynamical structure of the gluon plasma in the whole temperature range.

Here we summarize our main results, as follows:

- In the gluon plasma equation of state the confining dynamics at non-zero temperature is taken into account through the T-dependent bag constant and the mass gap Δ_R^2 itself. The gluon plasma pressure (9.3.39) compared to lattice pressure given in Ref. [221] is shown in Fig. 9.3.4. It demonstrates only a very small difference between them despite the completely different nature of analytical and lattice approaches to QCD at finite temperature.

- Other non-perturbative effects are also taken into account via the Yang–Mills and perturbative parts of the gluon plasma pressure, though the latter one is suppressed by the order of a small α_s.
- The mass gap Δ_R^2 or, equivalently, ω_{eff} and the asymptotic scale parameter for Yang–Mills fields Λ_{YM}^2 are the only two independent input scale parameters in our approach. They have not been introduced by hand, but they appear in the framework of the proposed formalism which is to be generalized to non-zero temperature and density.
- The presence of the two different types of massive gluonic excitations ω' and $\bar{\omega}$ with an effective masses 1.17 GeV and 0.585 GeV, respectively.
- The presence of the massive gluonic excitation $\alpha_s \cdot \bar{\omega}$ with the same effective mass 0.585 GeV, the propagation of which, however, is suppressed by the α_s-order. There is no place for some other types of massive gluonic excitations in our picture. The massive excitations of order $\alpha_s^2 \cdot \bar{\omega}$ and higher are, in principle, possible. However, numerically they are negligibly small.
- We have the two different types of massless gluonic excitations ω_1 and ω_2 which, nevertheless, are not free. All these massive and massless gluonic excitations are of the non-perturbative dynamical origin. They disappear from the gluon plasma spectrum in the $\Delta_R^2 = 0$ limit. So in total we have the five different massive and massless gluonic excitations in the gluon plasma.
- We also have the two different types of the excitations propagating only below T_c, which do not depend explicitly on the mass gap. Their propagations are described by the so-called Stefan–Boltzmann-type terms.
- The presence of the two different perturbative contributions above T_c. The Stefan–Boltzmann term $P_{SB}(T)$ itself and its asymptotic freedom correction $\alpha_s(T) \cdot P_{SB}(T)$.
- Below T_c all the independent thermodynamic quantities are exponentially suppressed in the $T \to 0$ limit, Figs. 9.3.5 and 9.8.1.
- Conformity and the velocity of sound squared, being the ratios of the corresponding independent thermodynamic quantities, demonstrate the non-trivial dependence on $T < T_c$, Figs. 9.8.3 and 9.8.4.
- The existence of the characteristic temperature $T_c = 266.5$ MeV at which the first-order phase transition occurs in the gluon plasma. The expo-

nential rise below T_c transforms to the polynomial one above it, but the gluon plasma pressure remains continuous at T_c, Figs. 9.3.4 and 9.3.5.

- The energy and entropy densities have finite jumps and $\epsilon_{LH} = 1.414$ at T_c, Fig. 9.3.5. The heat capacity has an essential discontinuity of a δ-type singularity at T_c, Fig. 9.8.1.
- The trace anomaly has a peak at T_c as plotted on Figs. 9.8.5, 9.8.6, and 9.8.7, and shows the highly non-trivial structure above T_c up to high temperature. It is not a simple polynomial-type in this region.
- Because of the presence of the mass gap term $\sim \alpha_s \Delta_R^2 T^2$ the trace anomaly goes down as $1/T^2$ in Fig. 9.8.5 and approaches the corresponding constants in Figs. 9.8.6 and 9.8.7 in the limit of very high temperatures.
- However, the trace anomaly has also the next-to-leading order correction $\sim (1/T^3)$ to the above-mentioned leading order term $\sim (1/T^2)$ at high temperatures. It comes from the non-analytical dependence on the mass gap in terms $\sim \Delta^3 T$ in this limit.
- Due to such a non-trivial dynamical structure of the trace anomaly, the gluon condensate gains rather small negative values at high temperatures, Fig. 9.8.8. It can be considered as one of the important constituents saturating the non-perturbative dynamical structure of the gluon plasma at finite temperatures.
- The non-perturbative physics of the mixed phase, in the temperature interval $(0.8 - 4)T_c$ in $SU(3)$ gluon plasma is well understood within our approach.
- Moreover, the behavior of $SU(3)$ gluon plasma in the region of low temperatures below $0.8T_c$ is also under control within our approach.
- Since the non-perturbative effects are still important above T_c, the behavior of $SU(3)$ gluon plasma in this region is different from the behavior of a gas of free massless gluons.
- All the independent thermodynamic quantities approach rather slowly their respective Stefan–Boltzmann limits at high temperatures, Figs. 9.3.5 and 9.8.1.
- The thermodynamic quantities which are the ratios of the corresponding independent counterparts rather rapidly approach their respective Stefan–Boltzmann limits at high temperatures, Figs. 9.8.2, 9.8.3 and 9.8.4.
- As a byproduct but necessary component of our calculations we have restored lattice pressure below $0.9T_c$, see Figs. 9.D.1 and 9.D.2.
- All analytical derivations agree with our numerical results and vice versa.

Our general conclusions are:

The non-perturbative analytic approach to thermal QCD provides a detailed numerical and dynamical description of $SU(3)$ gluon plasma in the whole temperature range, especially in the region of low temperatures below T_c, something that is not accessible for and is missing in the lattice approach to thermal QCD. In other words, for the first time we can numerically predict the behavior of all the thermodynamic quantities far below T_c. At the same time, our approach agrees very well with lattice calculations of the independent thermodynamic quantities at finite temperatures, though we have adjusted our values to those of lattice ones for the $SU(3)$ Yang – Mills pressure in Ref. [221] only for high temperatures far away from T_c. At low temperature all the independent thermodynamic quantities are exponentially suppressed. The dependence of all the thermodynamic quantities on the mass gap and the finite temperature is rather complicated. The polynomial regime for them appears only in the limit of high temperature. The approach of the independent thermodynamic quantities to their respective Stefan – Boltzmann limits is rather slow, while for their considered ratios it is rather rapid.

We report rather substantial deviations of our results from lattice ones for the trace anomaly plotted on Figs. 9.8.5, 9.8.6, and 9.8.7, which is very sensitive to the non-perturbative effects. The dynamical reason of these differences has to be traced back to the confining β-function in Appendix 9.B used in our calculations, which linearly depends on the mass gap squared. As repeatedly emphasized above, the β-functions used in all lattice simulations of thermal QCD are completely different, not depending on the mass gap at all.

We also predict the existence of the three massive and the two massless excitations, all of the non-perturbative dynamical origin. One of the massive excitations has an effective mass 1.17 GeV and the two others have the same effective mass 0.585 GeV, but propagating in different ways. Let us emphasize that the effective masses appear through the corresponding gluon mean numbers, and they are not the pole masses which appear in the corresponding propagators (see, for example [184, 234]). This means that we cannot assign to the corresponding massive excitations a meaning of being physical particles. They have to be treated rather as quasi-particles, indeed. The temperature-dependent gluon condensate is also present, playing an important role in the dynamical structure of the gluon plasma as well. The two different excitations of the so-called Stefan – Boltzmann-type (not explicitly depending on the mass gap) should exist in the gluon plasma below T_c, at least. The two pure perturbative massless gluonic excitations are necessary present above T_c as well. Their propagation is described by the Stefan – Boltzmann pressure and its asymptotic freedom correction.

In brief, now we already know what is the physics behind all the lattice curves and their numbers and even more beyond them. Let us underline that the gluon pressure $P_g(T)$ is the most important part of the gluon plasma pressure (9.3.39). Just it is mainly responsible for the non-perturbative physics described by the analytical Yang – Mills equation of state derived here. Without its correct analytical derivation and numerical calculation a firm agreement of the gluon plasma pressure (9.3.39) with lattice pressure in the whole temperature range would be impossible, see Fig. 9.3.4. As described in Section 9.3 in detail the addition of the perturbative contributions to $P_g(T)$ above T_c ensures the correct Stefan – Boltzmann limit of the gluon plasma pressure and its all derivatives at high temperature only. In other words, we reproduce all the lattice curves and numbers and deviation from them, for the trace anomaly, on general grounds. For this we need first of all the mass gap Δ_R^2 and then Λ_{YM}^2 and α_s correctly implemented into the used formalism at non-zero temperature.

Some other interesting finite temperature applications of our approach are: the solution of the relativistic hydrodynamics equations of motion [220, 228, 229], to calculate the corresponding transport coefficients, such as bulk and shear viscosities and its ratio to entropy density [248–250], etc. For these aims it suffices to use explicit expression for the gluon plasma pressure above T_c,

$$P(T) = [1 - \alpha_s(T)]\, P_{SB}(T) - 0.2815 P_g(T), \quad \text{at} \quad T > T_c, \qquad (9.10.1)$$

where we omit the next-to-leading term $\sim T^{-3}$ in equation (9.3.41). $\alpha_s(T)$ is given in equation (9.3.40). $P_g(T)$ is given by the sum of the corresponding integrals in Section 9.1. For the higher temperatures, one can use its high-temperature expansion,

$$P_g(T) \sim B_2 \alpha_s \Delta^2 T^2 + \left[B_3 \Delta^3 + M^3\right] T, \quad \text{at} \quad T \gg T_c, \qquad (9.10.2)$$

where the explicit expressions for both dimensionless constants B_2, B_3 and M^3, which has the dimension of the GeV^3, can be restored from the expansion (8.6.34).

We are also planning to include the quark degrees of freedom within our approach in order to derive the non-perturbative analytical equation of state for the quark–gluon plasma and compare it with already existing lattice quark–gluon plasma equation of state in Refs. [194–196, 251–253].

9.A Appendix: Analytical and numerical evaluation of the latent heat

It is instructive to calculate the auxiliary functions $L(T)$ and $H(T)$ starting from their general expressions (9.3.11) and (9.3.12). Of course, the final numerical results are consistent with their derived expressions shown in equation (9.6.9). From the expression (9.3.11) it follows that

$$\frac{\partial L(T)}{\partial T} = \frac{T_c}{T^2}\sum_{i=1}^{n}\alpha_i A_i e^{-\alpha_i (T_c/T)} P_{SB}(T) + \sum_{i=1}^{n} A_i e^{-\alpha_i (T_c/T)} \frac{\partial P_{SB}(T)}{\partial T} - \alpha\frac{T_c}{T^2} e^{-\alpha(T_c/T)} P_g(T) - e^{-\alpha(T_c/T)}\frac{\partial P_g(T)}{\partial T}, \tag{9.A.1}$$

and at $T = T_c$ it is

$$\left.\frac{\partial L(T)}{\partial T}\right|_{T_c} = \frac{1}{T_c}\sum_{i=1}^{n}\alpha_i A_i e^{-\alpha_i} P_{SB}(T_c) + \sum_{i=1}^{n} A_i e^{-\alpha_i} \left.\frac{\partial P_{SB}(T)}{\partial T}\right|_{T_c} - \frac{\alpha}{T_c} e^{-\alpha} P_g(T_c) - e^{-\alpha} \left.\frac{\partial P_g(T)}{\partial T}\right|_{T_c}. \tag{9.A.2}$$

In the same way from the expressions (7.5.1) and (9.3.41) it follows that

$$\frac{\partial H(T)}{\partial T} = (1 - \alpha_s(T))\frac{\partial P_{SB}(T)}{\partial T} + \frac{2.5047}{T}\left(\frac{T_c}{T}\right)^4 P_g(T) - \phi_h(T)\frac{\partial P_g(T)}{\partial T}, \tag{9.A.3}$$

where as pointed out above we have omitted the term $\alpha'_s(T) \sim -\alpha_s^2(T)$ due to the discussion in Section 9.7. At $T = T_c$ it is

$$\left.\frac{\partial H(T)}{\partial T}\right|_{T_c} = 0.64 \left.\frac{\partial P_{SB}(T)}{\partial T}\right|_{T_c} + \frac{2.5047}{T_c} P_g(T_c) - \phi_2(T_c) \left.\frac{\partial P_g(T)}{\partial T}\right|_{T_c}. \tag{9.A.4}$$

Taking further into account these relations and the relations (9.3.4) and (9.3.14), the latent heat given by equation (9.6.4) thus becomes

$$\epsilon_{LH}(T_c) = \left[2.5047 - 1.84\sum_{i=1}^{n}\alpha_i A_i e^{-\alpha_i} + \alpha e^{-\alpha}\right] P_g(T_c), \tag{9.A.5}$$

and in dimensionless units it is

$$\epsilon_{LH} \equiv \frac{\epsilon_{LH}(T_c)}{T_c^4} = \left[2.5047 - 1.84\sum_{i=i}^{n}\alpha_i A_i e^{-\alpha_i} + \alpha e^{-\alpha}\right] \cdot 0.9538, \tag{9.A.6}$$

where the number $P_g(T_c)/T_c^4$ has already been substituted. Using further the numbers (9.3.38), one finally obtains

$$\epsilon_{LH} = 1.414. \tag{9.A.7}$$

Let us remind that the terms depending on the parameters α_i and A_i in equation (9.A.5) are of the non-perturbative origin, but not explicitly depending on the mass gap. However, putting the fitting parameters formally zero $A_i = 0$, the latent heat, nevertheless, remains finite. Thus it is not a mere consequence of the fit for the pressure, especially knowing that the fit to lattice data has been done for the region far away from T_c. Since the fitting procedure is aiming to make the pressure continuous across T_c, the fitting parameters α_i and A_i will necessarily contribute to the numerical value of the gluon pressure's derivative at T_c. However, the numerical value for the latent heat in equation (9.A.7) really comes mainly from $P_g(T)$ and its derivative at T_c, which behave differently below and above T_c — it is clearly seen from the derivations in this Appendix, in general, and from equation (9.A.5), in particular. Let us remind once more that T_c at which the latent heat has been calculated is fixed again by $P_g(T)$ itself. Concluding, we did not use the lattice data for the energy density as well as for all other thermodynamic quantities, only for the pressure.

9.B Appendix: The β-function for the confining effective charge at non-zero temperature

Let us show explicitly the corresponding β-function for the intrinsically non-perturbative effective charge (4.4.1). From the renormalization group equation,

$$q^2 \frac{\mathrm{d}\alpha^{INP}(q^2;\Delta_R^2)}{\mathrm{d}q^2} = \beta\left(\alpha^{INP}\left(q^2;\Delta_R^2\right)\right), \tag{9.B.1}$$

it simply follows that

$$\beta\left(\alpha^{INP}\left(q^2;\Delta_R^2\right)\right) = -\alpha^{INP}\left(q^2;\Delta_R^2\right) = -\frac{\Delta_R^2}{q^2}. \tag{9.B.2}$$

Thus, the corresponding β-function as a function of its argument is always in the domain of attraction, thus negative. So it has no infrared stable fixed point indeed as it is required for the confining theory [2]. Let us remind that the confining effective charge (9.B.2), and hence its β-function, is a result of the summation of the non-perturbative, skeleton loop diagrams, contributing to the full gluon self-energy in the $q^2 \to 0$ regime —

the above-mentioned cluster expansion but in powers of the mass gap. This summation has been performed within the corresponding equations of motion.

In frequency–momentum space from equation (7.3.2) and (9.B.2) one gets

$$\beta^{INP}(\omega^2,\omega_n^2) = -\alpha^{INP}(\omega^2,\omega_n^2) = -\frac{\Delta_R^2}{\omega^2+\omega_n^2}. \tag{9.B.3}$$

The confining effective charge with the corresponding β-function (9.B.3) determines the structure of $SU(3)$ gluon plasma at low and finite temperatures (mixed phase) [215, 236] within our approach. In the limit of very high temperature its role is decreased, as expected.

9.C Appendix: Least Mean Squares method and the definition of the average deviation

The gluon plasma pressure $P_{GP}(T)$ at high temperatures above T_c, the right-hand-side of equation (9.3.29), depends on the parameters c_0, c and n. For a given n the parameters c_0 and c are determined by using the Least Mean Squares method [226]. This method makes it possible to calculate the values of the parameters c_0 and c for the last $p \geq 2$ number of data points available from lattice results. In this way the temperature region below $3.436657 \cdot T_c$ was covered with an approximation curve which best fits to lattice calculations according to this method. Let us note that for the accuracy of this method the number of data points p has to be sufficiently large.

However, the result of the calculations for c_0 and c depends on the number of data points p considered and the value for n, thus, $c_0 = c_0(p,n)$ and $c = c(p,n)$. In order to choose which value for n is preferable, we have introduced the average deviation $\Delta_p(n)$ for the given numbers of points p. For our purpose it is convenient to define this quantity as follows:

$$\Delta_p(n) = \frac{1}{p}\sum_{i=1}^{p}\frac{1}{T_i^4}\left|5.2634 \cdot P_l(T_i) - 3 \cdot P_{GP}(T_i,n)\right| . \tag{9.C.1}$$

Here $||$ denotes the absolute value and T_i is the temperature at the ith data point.

Our results are summarized in Fig. 9.C.1 for $n = 1,2,3$. We can conclude that the average deviation is minimal for $n = 3$ and at a sufficiently

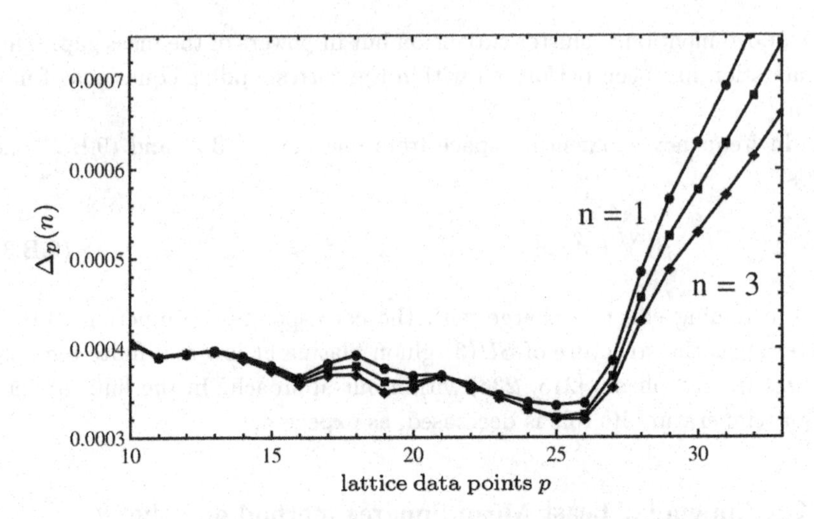

Fig. 9.C.1 Average deviations $\Delta_p(n)$ as functions of lattice data points p and $n = 1, 2, 3$.

large number of lattice data points $p = 26$. This makes it possible to finally fix $c_0 = 1.2815$ and $c = 2.1164$, see equation (9.3.30). For $n \geq 4$ the values of the coefficient c do not satisfy the restriction (9.3.28), they become bigger than its upper bound 2.1776.

9.D Appendix: Restoration of the lattice pressure below $0.9T_c$

The lattice pressure below T_c calculated in [221] is shown in Fig. 9.D.1. It goes down to zero at $0.8T_c$, more precisely there are no convincing lattice data points below this value. As we have mentioned above this is due to large uncertainties in lattice simulations at low temperatures. Our aim here is to restore the true lattice pressure below $0.9T_c$ by reproducing lattice pressure in the region $(0.907850 - 0.989528) \cdot T_c$, where it is considered to be correct. Using the Least Mean Squares method within our approach we have reproduced very well the lattice pressure above T_c and at $T = T_c$.

Let us remind that there is a very small difference between our value and the lattice value at this point, see equations (9.3.34) and (9.3.35). On the one hand, this means that we can trust lattice pressure in the above-mentioned region. On the other hand, this means that we can expect that our method to calculate both pressures below T_c will be as good as it was in reproducing the lattice pressure above $T \geq T_c$, see Fig. 9.3.2. However,

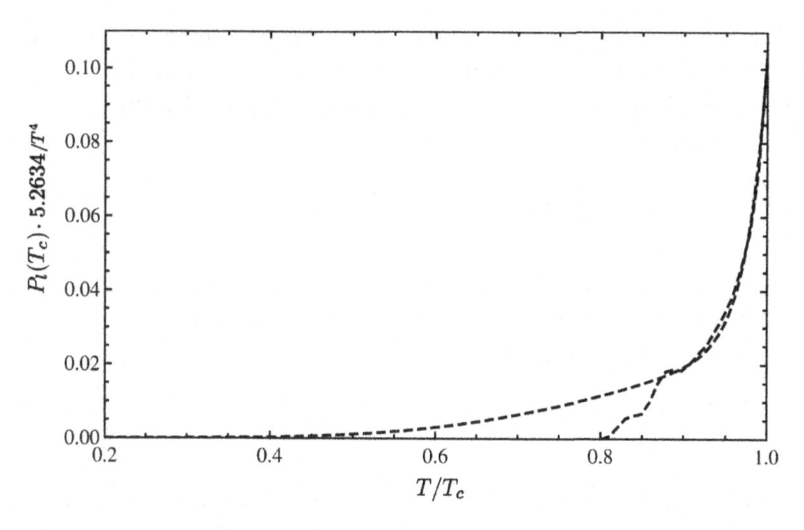

Fig. 9.D.1 Restored lattice pressure (9.D.4) as a function of T/T_c are shown below $T = 0.9T_c$. Above this value it reproduces lattice pressure in Ref. [221].

within our method and for our purpose we need to know the lattice value for the parameter $\phi_h(1) = \phi_h(T_c)$, since it enters the relation (9.3.14) and equation (9.3.15). Numerically it can be found by adjusting the left-hand-side of equation (9.3.15) to the lattice data in Ref. [221] with the help of equation (9.3.19) as follows:

$$\frac{P_l(T_c)}{T_c^4} \cdot 5.2634 = [2.1776 - \phi_h(T_c)] \, \frac{3P_g(T_c)}{T_c^4}, \tag{9.D.1}$$

where all numbers are known, apart from $\phi_h(T_c)$ itself. Substituting lattice data as described above, i.e., using $P_l(T_c)/T_c^4 = 0.019676$ and our value for $3P_g(T_c)/T_c^4 = 2.8616$, one obtains

$$\phi_h^l(1) = \phi_h^l(T_c) = 2.1415, \tag{9.D.2}$$

where superscript the 'l' shows that this is the lattice value. Comparing this value with the value in (9.3.31), one concludes that there is a small difference between them (in the second digit after point, indeed).

Substituting this value into the equation (9.3.14), it finally becomes

$$\sum_{i=1}^{n} A_i e^{-\alpha_i} = \left[0.5435e^{-\alpha} - 0.5238\right], \tag{9.D.3}$$

which makes it possible to reduce the number of independent parameters from $2n + 1$ to $2n$.

The fit to lattice data available from the low temperature interval in [221], namely $a = 0.907850 - 0.989528$, is to be performed with the help of the following equation — the value $a = 1$ has been already used in order to derive the relation (9.D.3):

$$\frac{P_l(T)}{T^4} \cdot 5.2634 = 5.2634 \cdot \sum_{i=1}^{n} A_i e^{-\alpha_i \frac{T_c}{T}} + \left(1 - e^{-\alpha \frac{T_c}{T}}\right) \frac{3P_g(T)}{T^4}, \quad (9.D.4)$$

where $T = aT_c$ and the values of $P_l(T)/T^4$ have been taken from lattice data, while the values of $3P_g(T)/T^4$ have been taken from our data, Fig. 9.2.1. The best fit has been achieved at

$$\alpha = 0.0001, \ \alpha_1 = 39.6, \ \alpha_2 = 3.4, \ A_1 e^{-\alpha_1} = 0.0146, \text{ and } A_2 e^{-\alpha_2} = 0.005. \quad (9.D.5)$$

This fit is shown in Fig. 9.D.1. On the one hand, these values show that the deep penetration of the Stefan – Boltzmann-type terms below T_c is indeed strongly suppressed, as expected. On the other hand, they satisfy the relation (9.D.3) as it should be. Possibly the fit can be slightly improved by taking into account more terms in the left-hand-side of the relation (9.D.3). We restrict ourselves to the two terms only, since the values of the lattice

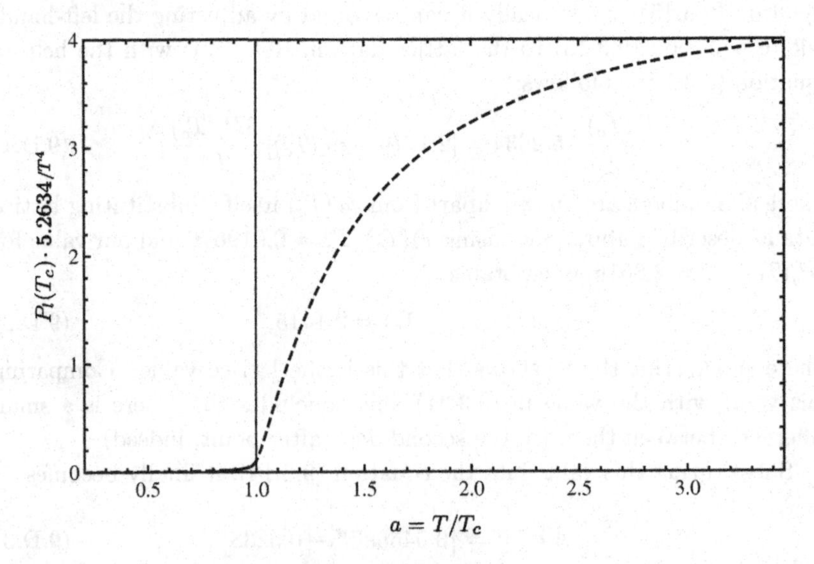

Fig. 9.D.2 Restored lattice pressure (9.D.4) as a function of T/T_c is shown in the whole temperature interval $a = T/T_c$. Above $T = 0.9T_c$ it reproduces lattice pressure in Ref. [221].

pressure below T_c are very small and there are no convincing lattice data points below $0.9T_c$. Let us also note that we do not use the Least Mean Squares method here, since we have encountered some numerical problems with its non-linear realization. However, our fit made by 'hand' is very accurate for the above-mentioned considered interval.

We are able now to establish the behavior of the lattice pressure in the temperature interval $a \leq 1$ as well — for convenience, in Fig. 9.D.1 it is shown up to $0.2T_c$ only. This should be achieved with the help of the same equation (9.D.4), but where all the numbers are already known from the values (9.D.5). Thus there is no doubt that we can reproduce the lattice curve almost identically also very close below T_c. Therefore, we can consider the curve below $0.9T_c$ in Fig. 9.D.1 as a true lattice curve in this region. The restored lattice curve in the whole temperature range in shown in Fig. 9.D.2.

Problems

Problem 9.1. *Derive eqaution* (9.3.8) *from the expression given in Ref.* [193].

Problem 9.2. *Derive all equations in Section 9.6.*

Problem 9.3. *Derive all equations in Section 9.7.*

Problem 9.4. *Derive equation* (9.9.1).

Problem 9.5. *Derive all equations in Appendix 9.A.*

Bibliography

[1] H. Fritzsch, M. Gell-Mann, and H. Leutwyler, Phys. Lett. B, **47** (1973) 365.

[2] W. Marciano and H. Pagels, Phys. Rep. C, **36** (1978) 137.

[3] M.E. Peskin and D.V. Schroeder, *An Introduction to Quantum Field Theory* (AW, Advanced Book Program, 1995).

[4] C. Itzykson and J.-B. Zuber, *Quantum Field Theory*, (Mc Graw-Hill Book Company, 1984).

[5] T. Muta, *Foundations of QCD* (Word Scientific, 1987).

[6] R.D. Field, *Application of Perturbative QCD* (Addison-Wesley, 1990).

[7] A.S. Kronfeld and C. Quigg, arXiv:1002.5032 (2010).

[8] G. 't Hooft, *Conference on Lagrangian Field Theory*, Marseille, 1972.

[9] D.J. Gross and F. Wilczek, Phys. Rev. Lett., **30** (1973) 1343.

[10] H.D. Politzer, Phys. Rev. Lett., **30** (1973) 1346.

[11] A. Jaffe and E. Witten, Yang-Mills Existence and the Mass Gap, http://www.claymath.org/prize-problems/, http://www.arthurjaffe.com

[12] G. 't Hooft, arXiv:hep-th/0408183 (2004).

[13] A.M. Polyakov, arXiv:hep-th/0407209 (2004).

[14] E.G. Eichtein and F.L. Feinberg, Phys. Rev. D, **10** (1974) 3254.

[15] M. Baker and C. Lee, Phys. Rev. D, **15** (1977) 2201.

[16] U. Bar-Gadda, Nucl. Phys. B, **163** (1980) 312.

[17] R. Alkofer and L. von Smekal, Phys. Rep., **353** (2001) 281.

[18] C.D. Roberts and A.G. Williams, Prog. Part. Nucl. Phys., **33** (1994) 477.

[19] P. Maris and C.D. Roberts, Int. J. Mod. Phys. E, **12** (2003) 297.

[20] J.C. Taylor, Nucl. Phys. B, **33** (1971) 436.

[21] A.A. Slavnov, Sov. Jour. Theor. Math. Phys., **10** (1972) 153.

[22] G. 't Hooft, Nucl. Phys. B, **33** (1971) 173.

[23] S.K. Kim and M. Baker, Nucl. Phys. B, **164** (1980) 152.

[24] S.-H.H. Tye, E. Tomboulis, and E.C. Poggio, Phys. Rev. D, **11** (1975) 2839.

[25] P. Pascual, R. Tarrach, Nucl. Phys. B, **174** (1980) 123.

[26] B.W. Lee and Phys. Rev. D, **9** (1974) 933.

[27] V. Gogokhia, Phys. Lett. B, **618** (2005) 103.

[28] V.N. Gribov and J. Nyiri, *Quantum Electrodynamics* (Cambridge University Press, 2001).
[29] G. 't Hooft and M. Veltman, Nucl. Phys. B, **44** (1972) 189.
[30] V. Gogokhia and Gy. Kluge, Phys. Rev. D, **66** (2002) 056013.
[31] V. Gogohia, Phys. Lett. B, **531** (2002) 321.
[32] V. Gogohia, Gy. Kluge, and I. Vargas de Usera, Phys. Lett. B, **576** (2003) 243.
[33] R. Feynman, Nucl. Phys. B, **188** (1981) 479.
[34] H. Pagels, Phys. Rev. D, **15** (1977) 2991.
[35] V.Sh. Gogokhia, Phys. Rev. D, **41** (1990) 3279.
[36] V. Gogokhia, Int. J. Theor. Phys., **48** (2009) 3061.
[37] V. Gogokhia, Int. J. Theor. Phys., **48** (2009) 3470.
[38] V. Gogokhia, Int. J. Theor. Phys., **48** (2009) 3449.
[39] V. Gogohia, Phys. Lett. B, **584** (2004) 225.
[40] I.M. Gelfand and G.E. Shilov, *Generalized Functions, vols. I, II* (Academic Press, NY, 1968).
[41] V. Gogohia and Gy. Kluge, Phys. Lett. B, **477** (2000) 387.
[42] V. Gogohia and M. Prisznyák, Phys. Lett. B, **494** (2000) 109.
[43] J.M. Cornwall, Phys. Rev. D, **26** (1982) 1453.
[44] V.A. Rubakov, *Classical Gauge Fields* (Editorial YRSS, Moscaw, 1999).
[45] L. von Smekal, A. Hauk, and R. Alkofer, Ann. Phys., **267** (1998) 1.
[46] R. Alkofer and J. Greensite, J. Phys. G, **34** (2007) S3.
[47] A.C. Aguilar, D. Binosi, J. Papavassiliou, Phys. Rev. D, **78** (2008) 025010.
[48] C.S. Fischer, A. Maas, and J.H. Pawlowski, arXiv:0810.1987 (2008).
[49] S.P. Sorella, arXiv:0905.1010 (2009).
[50] D. Dudal, J.A. Gracey, S.P. Sorella, N. Vandersickel, and H. Verschelde, Phys. Rev. D, **78** (2008) 065047.
[51] D. Zwanziger, arXiv:0904.2380 (2009).
[52] L. Baulieu, M.A.L. Capri, A.J. Gomez, V.E.R. Lemes, R.F. Sobreiro, and S.P. Sorella, Eur. Phys. J. C **66** (2010) 451.
[53] K.-I. Kondo, arXiv:0907.3249 (2009).
[54] M.A. Lavrentiev and B.V. Shabat, *Methods of the theory of functions of complex variable*, (Russ. Ed., Moscow, Nauka, 1987).
[55] A.I. Alekseev and B.A. Arbuzov, Phys. Atom. Nucl., **61** (1998) 264.
[56] K.D. Born *et al.*, Phys. Lett. B, **329** (1994) 325.
[57] V.M. Miller *et al.*, Phys. Lett. B, **335** (1994) 71.
[58] V. Gogokhia, arXiv:0704.3189.
[59] V. Gogohia and Gy. Kluge, Phys. Rev. D, **62** (2000) 076008.
[60] V. Gogohia *et al.*, Int. Jour. Mod. Phys. A, **15** (2000) 45.
[61] G.G. Barnaföldi and V. Gogokhia, J. Phys. G, **37** (2010) 025003.
[62] V. Gogohia, arXiv:hep-ph/0311061 (2003).
[63] S. Mandelstam, Phys. Rev. D, **20** (1979) 3223.
[64] V.N. Gribov, Nucl. Phys. B **139** (1978) 1.
[65] V.N. Gribov, *Gauge Theories and Quark Confinement* (PHASIS, Moscow, 2002).
[66] Y.L. Dokshitzer and D.E. Kharzeev, arXiv:hep-ph/0404216 (2004).

[67] G. Preparata, Phys. Rev. D, **7** (1973) 2973.
[68] V. Gogohia, Gy. Kluge, and M. Prisznyák, hep-ph/9509427. (1995)
[69] G. Pocsik and T. Torma, Acta Phys. Hun., **62** (1987) 101-107.
[70] Z. Fodor, Acta Phys. Pol., B **19** (1988) 21.
[71] G. 't Hooft, Nucl. Phys. B, **75** (1974) 461.
[72] K.B. Wilson, Phys. Rev. D, **10** (1974) 2445.
[73] M. Bander, Phys. Rep., **75** (1981) 205.
[74] *Quark Matter 2005*, Edited by T. Csorgo, G. David, P. Lévai, and G. Papp (Elsevier, Amsterdam-...-St. Louis, 2005).
[75] M. Gyulassy and L. McLerran, arXiv:nucl-th/0405013 (2004).
[76] A.S. Kronfeld, arXiv:1007.1444 (2010).
[77] A.S. Kronfeld, arXiv:1203.1204, to appear in Ann. Rev. Nucl. Part. Science. (2013).
[78] S. Adler and A.C. Davis, Nucl. Phys. B, **244** (1984) 469.
[79] V.Sh. Gogokhia and B.A. Magradze, Phys. Lett. B, **217** (1989) 162.
[80] V. Gogokhia and B. Magradze, Mod. Phys. Lett. A, **4** (1989) 1549.
[81] V.Sh. Gogokhia, Phys. Lett. B, **224** (1989) 177.
[82] V.Sh. Gogokhia, Phys. Rev. D, **40** (1989) 4157.
[83] M. Cheng *et al.*, Phys. Rev. D **77** (2008) 014511.
[84] A. Bazazov *et al.*, Phys. Rev. D, **80** (2009) 014504.
[85] J. Zimányi, P. Lévai, and T.S. Biró, arXiv:hep-ph/0205192 (2002).
[86] T.S. Biró, P. Lévai, and J. Zimányi, J. Phys. G, **28** (2002) 1561.
[87] Z. Fodor and S.D. Katz, arXiv:0908.3341 (2009).
[88] I.V. Andreev, *QCD and hard processes at high energies* (Moscaw, Nauka, 1981).
[89] F. Wilczek, *Proc. Inter, Conf., QCD–20 Years Later, v. 1* (Aachen, June 9–13, 1992).
[90] *Confinement, Duality, and Nonperturbative Aspects of QCD*, edited by P. van Baal, NATO ASI Series B: Physics, vol. **368** (Plenum, New York, 1997).
[91] *Non-Perturbative QCD, Structure of the QCD Vacuum*, edited by K-I. Aoki, O. Miymura, and T. Suzuki, Prog. Theor. Phys. Suppl., **131** (1998) 1.
[92] Mark D. Roberts, arXiv:hep-th/0012062.
[93] P. van Baal, Nucl. Phys. B, **63A-C** (1998) 126.
[94] F. Lenz, J.W. Negele, and M. Thies, Phys. Rev. D, **69** (2004) 074009.
[95] G. Gabadadze, M. Shifman, arXiv:hep-ph/0206123.
[96] N.K. Nielsen and P. Olesen, Nucl. Phys. B, **144** (1978) 376.
[97] L. Del Debbio, M. Faber, J. Giedt, J. Greensite, and S. Olejnik, Phys. Rev. D, **58** (1998) 094501.
[98] M. Engelhardt and H. Reinhardt, Nucl. Phys. B, **585** (2000) 591.
[99] G. Ripka, *Lectures Notes in Physics (LNP)*, (Springer, Verlag, v. 639, 2004).
[100] A.A. Belavin, A.M. Polyakov, A. Schwartz, and Y. Tyupkin, Phys. Lett. B, **59** (1975) 85.
[101] T. Schafer and E.V. Shuyrak, Rev. Mod. Phys. **70** (1998) 323.
[102] V. Gogohia, *et al.*, Phys. Lett. B, **453** (1999) 281.
[103] V. Gogohia, Phys. Lett. B, **485** (2000) 162.

[104] S. Weinberg, Phys. Rev. Lett. **31** (1973) 494.
[105] H. Georgi and S. Glashow, Phys. Rev. Lett. **32** (1974) 438.
[106] J.L. Gervais and A. Neveu, Phys. Rep. C, **23** (1976) 240.
[107] L. Susskind and J. Kogut, Phys. Rep. C, **23** (1976) 348.
[108] S.G. Matinyan and G.K. Savvidy, Nucl. Phys. B, **134** (1978) 539.
[109] M. Baker, J. S. Ball, and F. Zachariasen, Nucl. Phys. B, **186** (1981) 531.
[110] N. Brown and M. R. Pennington, Phys. Rev. D, **39** (1989) 2723.
[111] D. Atkinson, *et al.*, J. Math. Phys. **25** (1984) 2095.
[112] L.G. Vachnadze, N.A. Kiknadze, and A.A. Khelashvili, Sov. Jour. Teor. Math. Phys. **102** (1995) 47.
[113] A.I. Alekseev and B. A. Arbuzov, Mod. Phys. Lett. A **13** (1998) 1747.
[114] X.-H. Guo and T. Huang, Il Nuovo Cim. A, **110** (1997) 799.
[115] T.H. Phat and N.T. Anh, Il Nuovo Cim. A, **110** (1997) 337.
[116] A. Hadicke, Intr. Jour. Mod. Phys. A, **6** (1991) 3321.
[117] W.J. Schoenmaker, Nucl. Phys. B, **194** (1982) 535.
[118] E.J. Gardner, J. Phys. G: Nucl. Phys., **9** (1983) 139.
[119] C. Michael, Nucl. Phys. B (Proc. Suppl.) **42** (1995) 147.
[120] G. Damm, W. Kerler, and V.K. Mitrjushkin, Phys. Lett. B, **443** (1998) 88.
[121] G. Burgio, F. Di Renzo, C. Parrinello, and C. Pittori, Nucl. Phys. B, **73** (1999) 623.
[122] G. Burgio, F. Di Renzo, G. Marchesini, and E. Onofri, Nucl. Phys. B, **63A-C** (1998) 805.
[123] M. Lusher **et al.**, Nucl. Phys. B, **413** (1994) 481.
[124] A.V. Nesterenko, Phys. Rev. D, **64** (2001) 116009.
[125] V.P. Nair and C. Rosenzweig, Phys. Rev. D, **31** (1985) 401.
[126] H. Narnhofer and W. Thirring, Springer Tracks Mod. Phys., **119** (1990) 1.
[127] K. Johnson, L. Lellouch, and J. Polonyi, Nucl. Phys. B, **367** (1991) 675.
[128] U. Ellwanger, M. Hirsch, and A. Weber, Eur. Phys. Jour. C, **1** (1998) 563.
[129] D.-U. Jingnickel and C. Wetterich, arXiv:hep-ph/0710397 (2001).
[130] H.J. Munczek, Phys. Lett. B, **175** (1986) 215.
[131] D.W. McKay and H.J. Munczek, Phys. Rev. D, **55** (1997) 2455.
[132] V. Gogohia, Gy. Kluge, and J. Nyiri, arXiv:hep-ph/0204347 (2002).
[133] S. Coleman and E. Weinberg, Phys. Rev. D, **7** (1973) 1888.
[134] D.J. Gross and A. Neveu, Phys. Rev. D, **10** (1974) 3235.
[135] M.A. Shifman, A.I. Vainshtein, and V.I. Zakharov, Nucl. Phys. B, **147** (1979) 385.
[136] V.A. Novikov, M.A. Shifman, A.I. Vainshtein, and V.I. Zakharov, Nucl. Phys. B, **191** (1981) 301.
[137] I. Halperin and A. Zhitnitsky, Nucl. Phys. B, **539** (1999) 166.
[138] E. Witten, Nucl. Phys. B, **156** (1979) 269.
[139] G. Veneziano, Nucl. Phys. B, **159** (1979) 213.
[140] V. Gogohia, Phys. Lett. B, **501** (2001) 60.
[141] M. Gell-Mann, R.J. Oakes, and B. Renner, Phys. Rev., **175** (1968) 2195.
[142] M. Creutz, *Quarks, Gluons and Lattice* (Cambridge, 1883)
[143] P. Hasenfratz, Prog. Theor. Phys. Suppl., **131** (1998) 189.

[144] M.-C. Chu, J.M. Grandy, S. Huang, and J.W. Negele, Phys. Rev. D, **49** (1994) 6039.
[145] T. DeGrand, A. Hasenfratz, T.G. Kovács, Nucl. Phys. B, **505** (1997) 417.
[146] A.S. Kronfeld, arXiv:hep-ph/0209321, Jour. ref.: eConfCO20620:FRBT05, (2002).
[147] A.M. Polyakov, Nucl. Phys. B, **486** (1997) 23.
[148] V.I. Zakharov, Int. Jour. Mod. Phys. A, **14** (1999) 4865.
[149] A.I. Alekseev and B.A. Arbuzov, arXiv:hep-ph/0407056, / arXiv:hep-ph/0411339 (2004).
[150] R. Streater and A. Wightman, *Spin and Statistics and all That* (W.A. Benjamin, NY 1964).
[151] F. Strocchi, Phys. Lett. B, **62** (1976) 60.
[152] A. Chodos *et al.*, Phys. Rev. D, **9** (1974) 3471.
[153] T. DeGrand *et al.*, Phys. Rev. D, **12** (1975) 2060.
[154] E.V. Shuryak, Phys. Rep., **115** (1984) 151.
[155] M.S. Chanowitz and S. Sharpe, Nucl. Phys. B, **222** (1983) 211.
[156] J.M. Cornwall, R. Jackiw, and E. Tomboulis, Phys. Rev. D, **10** (1974) 2428.
[157] A. Barducci *et al.*, Phys. Rev. D, **38** (1988) 238.
[158] R.W. Haymaker, Riv. Nuovo Cim., **14** (1991) 1.
[159] R.J. Crewther, Phys. Rev. Lett., **28** (1972) 1421.
[160] M.S. Chanowitz and J. Ellis, Phys. Rev., D **7** (1973) 2490.
[161] J.C. Collins, A. Duncan, and S.D. Joglecar, Phys. Rev. D, **16** (1977) 438.
[162] B. Grinstein and L. Randall, Phys. Lett. B, **217** (1989) 335.
[163] S. Narison, Phys. Lett. B, **387** (1996) 162.
[164] P. Castorina and So-Y. Pi, Phys. Rev. D, **31** (1985) 411.
[165] M.N. Chernodub, M.I. Polikarpov, and V.I. Zakharov, Phys. Lett. B, **457** (1999) 147.
[166] V. Gogohia, H. Toki, T. Sakai, and Gy. Kluge, Inter. Jour. Mod. Phys. A, **15** (2000) 45.
[167] Particle data group, Jour. Phys. G: Nucl. Part. Phys., **33** (2006) 1.
[168] Statistical Office of the European Communities, *http://epp.eurostat.ec. europa.eu*
[169] Energy for the Future — The Nuclear Option, A position paper of the EPS, *www.eps.org*
[170] V. Gogokhia, arXiv:hep-ph/0508224 (2005).
[171] F.D. Steffen, Eur. Phys. J. C, **59** (2009) 557.
[172] M.K. Gaillard and B. Zumino, Eur. Phys. J. C, **59** (2009) 213.
[173] S. Weinberg, UTTG-10-96, arXiv:astro-ph/9610044 (1996).
[174] C. Quigg, FERMILAB-PUB-09/230-T, arXiv:0905.3187 (2009).
[175] E. Komatsu *et al.* [WMAP Collab.], Astrophys. J. Suppl. **180** (2009) 330.
[176] F.R. Urban and A.R. Zhitnitsky, arXiv:0909.2684 (2009).
[177] K.A. Milton, arXiv:hep-th/0406024 (2004).
[178] L.H. Ford, arXiv:0911.3597 (2009).
[179] G.L. Klimchitskaya and V.M. Mostepanenko, arXiv:quant-ph/0609145 (2006).
[180] S.-S. Xue, Phys. Lett. B, **508** (2001) 211.

[181] S.-S. Xue, Phys. Rev. D, **68** (2003) 013004.
[182] M. Scandurra, arXiv:hep-th/0104127 (2001).
[183] J.D. Bjorken and S.D. Drell, *Relativistic Quantum Mechanics* (McGraw-Hill Book Company, 1978).
[184] J.I. Kapusta and C. Gale, *Finite-Temperature Field Theory* (Cambridge University Press, 2006).
[185] V. Gogohia and Gy. Kluge, Phys. Lett. B, **477** (2000) 387.
[186] V. Gogohia and M. Prisznyák, Phys. Lett. B **494** (2000) 109.
[187] L. McLerran, J. Phys. G, **35** (2008) 104001.
[188] R.M. Weiner, Int. Jour. Mod. Phys. E, **15** (2006) 37.
[189] L.P. Csernai *Introduction to relativistic Heavy Ion Collisions* (J. Wiley and Sons, 1994).
[190] *Proc. Quark Matter* J. Phys. G, **38** 11 Y. Schutz and U. Wiedemann (eds.) (2011).
[191] K. Kajantie, M. Lane, K. Rummukainen, and Y. Schroder, Phys. Rev. D, **67** (2003) 105008.
[192] E. Braaten and R.D. Pisarski, Nucl. Phys. B, **337** (1990) 569.
[193] J. Letessier and J. Rafelski, Phys. Rev. C, **67** (2003) 031902.
[194] Y. Aoki, Z. Fodor, S.D. Katz, and K.K. Szabó, J. High Energy Phys., **0601** (2006) 089.
[195] C. Schmidt, Z. Fodor, and S.D. Katz, PoS LAT2005, (2005) 163.
[196] F. Karsch, J. Phys. G, **35** (2008) 104096.
[197] M. Cheng *et al.*, Phys. Rev. D, **77** (2008) 014511.
[198] R. Gupta *et al.*, PoS LATTICE2008, (2008) 170.
[199] P. Lévai and U. Heinz, Phys. Rev. C, **57** (1998) 1879.
[200] T.S. Biró, A.A. Shanenko, and V.D. Toneev, Physics of Atomic Nuclei, **66** (2003) 982.
[201] K.K. Szabó and A.I. Tóth, J. High Energy Phys., **0306** (2003) 008.
[202] A. Peshier, B. Kämpfer, O.P. Pavlenko, and G. Soff, Phys. Rev. D, **54** (1996) 2399.
[203] A. Peshier, B. Kämpfer, and G. Soff, Phys. Rev. C, **61** (2000) 045203.
[204] M. Bluhm, B. Kämpfer, and G. Soff, Phys. Lett. B, **620** (2005) 131.
[205] M.A. Thaler, R.A. Schneider, and W. Weise, Phys. Rev. C, **69** (2004) 035210.
[206] C. Ratti, S. Roessner, M. A. Thaler, and W. Weise, Eur. Phys. J. C, **49** (2007) 213
[207] Y.B. Ivanov, V.V. Skokov, and V.D. Toneev, Phys. Rev. D, **71** (2005) 014005
[208] W. Cassing, Nucl. Phys. A, **791** (2007) 365.
[209] A. Leonidov, K. Redlich, H. Satz, E. Subonen, and G. Weber, Phys. Rev. D, **50** (1994) 4657.
[210] T. Schäfer, Nucl. Phys. B, **575** (2000) 269.
[211] Ph. Boucaud, J.P. Leroy, J. Micheli, O. Péne, and C. Roiesnel, arXiv:hep-ph/9810437 (1998).
[212] L. Dolan and R. Jakiw, Phys. Rev. D, **9** (1974) 3320.
[213] A. Rebhan, hep-ph/0105183 (2001).

[214] I.S. Gradshteyn and I.M. Ryzhik, *Tables of Integrals, Series, and Products* (Academic Press, 2007).
[215] V. Gogokhia and M. Vasúth, J. Phys. G, **37** (2010) 075015.
[216] J.-P. Blaizot and E. Iancu, Phys. Rep., **359** (2002) 355.
[217] A.P. Prudnikov, Y.A. Brichkov, and O.I. Marichev, *Integrals and Series* (Moscow, "NAUKA", 1981).
[218] P.N. Meisinger, T.R. Miller, and M.C. Ogilvie, Phys. Rev. D, **65** (2002) 034009.
[219] R.D. Pisarski, Prog. Theor. Phys. Suppl., **168** (2007) 276.
[220] E.V. Shuryak, Prog. Part. Nucl. Phys., **62** (2009) 48.
[221] M. Panero, Phys. Rev. Lett., **103** (2009) 232001.
[222] E. Megías, E.R. Arriola, and L.L. Salcedo, Phys. Rev. D, **80** (2009) 056005.
[223] T.S. Biró and J. Cleymans, Phys. Rev. C, **78** (2008) 034902.
[224] G. Boyd, J. Engels, F. Karsch, E. Laermann, C. Legeland, M. Lutgemeier, and B. Petersson, Nucl. Phys. B, **469** (1996) 419.
[225] F. Flechsig, A.K. Rebhan, and H. Schulz, Phys. Rev. D, **52** (1995) 2994.
[226] I.N. Bronshtein, K.A. Semendyayev, G. Musiol, and H. Muehlig, *Handbook of Mathematics* (Springer, 2004).
[227] H. Leutwyler, *Proceedings of the Conference QCD — 20 years later*, (ed. by P.M. Zerwas and H.A. Kastrup, Singapore: World Scientific, 1993) 693.
[228] T.S. Biró, *Is There a Temperature?* (Fundamental Theories of Physics 1014, Springer, 2011).
[229] M.I. Nagy, T. Csörgő, and M. Csanád, Phys. Rev. C, **77** (2008) 024908.
[230] B. Lucini, M. Teper, and U. Wenger, J. High Enery Phys., **02** (2005) 003.
[231] S. Datta and S. Gupta, Phys. Rev. D, **82** (2010) 114505.
[232] V. Mathieu, A.K. Kochelev, and V. Vento, Int. J. Mod. Phys. E, **18** (2009) 1.
[233] D.H. Rischke, M.I. Gorenstain, A. Schäfer, H. Stöcker and W. Greiner, Phys. Lett. B **278**, (1992) 19.
[234] J.-P. Blaizot and E. Iancu, Nucl. Phys. B, **459** (1995) 559.
[235] V. Gogokhia and M. Vasúth, arXiv:1012.4157 (2010).
[236] M. Gyulassy and L. McLerran, Nucl. Phys. A, **750**, (2005) 30.
[237] S. Weinberg, *The Quantum Theory of Fields, v. I* (Cambridge University Press, 1995).
[238] J.S. Ball, T.-W. Chiu, Phys. Rev. D, **22** (1980) 2542.
[239] G. Leibbrandt, *Noncovariant Gauges* (WS, Singapore, 1994).
[240] A. Bassetto, G. Nardelli, and R. Soldati, *Yang-Mills Theories in Algebraic Non Covariant Gauges* (WS, Singapore, 1991).
[241] E.R. Speer, Jour. Math. Phys., **15** (1974) 1.
[242] N.N. Bogoliubov and D.V. Shirkov, *Introduction to the Theory of Quantised Fields* (Interscience Publisher, NY, 1959).
[243] A. Cucchieri and T. Mendes, arXiv:0904.4033 (2009).
[244] A. Cucchieri and T. Mendes, Phys. Rev. Lett., **100** (2008) 241601.
[245] V.G. Bornyakov, V.K. Mitrjushkin, and M. Müller-Preussker, arXiv:0812.2761 (2008).

[246] I.L. Bogolubsky, E.-M. Igenfritz, M. Müller-Preussker, and A. Sternbeck, Phys. Lett. B, **676** (2009) 69.
[247] Ph. Boucaud, J.P. Leroy, A. Le Yaouanc, J. Micheli, O. Péne, and J. Rodríguez-Quintero, JHEP 0806 (2008) 099.
[248] P. Arnold, G.D. Moore, and L.G. Yaffe, JHEP (2000) 0011:001.
[249] P. Arnold, G.D. Moore, and L.G. Yaffe, JHEP (2003) 0305:051.
[250] A. Jakovác, arXiv:0911.3248v2 (2009).
[251] C. Bernard *et al.*, J. Phys. CS, **69** (2007) 012029.
[252] S. Borsányi *et al.*, J. Phys. G, **38** (2011) 124101.
[253] P. Huovinen and P. Petreczky, J. Phys. G, **38** (2011) 124103.

Index